Telecommunications
Pocket Reference

OTHER McGRAW-HILL TELECOMMUNICATIONS BOOKS OF INTEREST

Telecommunications Pocket Reference

Travis Russell

McGRAW-HILL
New York   San Francisco   Washington, D.C.   Auckland   Bogotá
Caracas   Lisbon   London   Madrid   Mexico City   Milan
Montreal   New Delhi   San Juan   Singapore
Sydney   Tokyo   Toronto

McGraw-Hill

A Division of The McGraw-Hill Companies

1 2 3 4 5 6 7 8 9 0   DOC/DOC   0 6 5 4 3 2 1 0

ISBN 0-07-135140-X

The sponsoring editor for this book was Steve Chapman, the editing supervisor was Frank Kotowski, Jr., and the production supervisor was Pamela A. Pelton.

It was set in Times by Paul Scozzari and Deirdre Sheean of McGraw-Hill's Professional Book Group Composition Unit in Hightstown, N.J.

Printed and bound by R. R. Donnelley and Sons, Inc.

McGraw-Hill books are available at special quantity discounts to use as premiums and sales promotions, or for use in corporate training programs. For more information, please write to the Director of Special Sales, McGraw-Hill, Two Penn Plaza, New York, NY 10121-2298. Or contact your local bookstore.

This book is printed on recycled, acid-free paper containing a minimum of 50% recycled, de-inked fiber.

IN MEMORY OF T. W. MILLER,
WHO ALWAYS GAVE LOVE AND LAUGHTER
IN THE MIDST OF HIS OWN PAIN AND SUFFERING.

Contents

CHAPTER 10. INTEGRATED SERVICES DIGITAL NETWORK (ISDN)   **201**

Acknowledgments

First, I must thank my editor Steve Chapman for coming up with the idea for this pocket guide. I'll have to find some shirts with bigger pockets, though. I would also like to thank Tekelec, Inc., for their ongoing support in this and all my book projects. Finally, many thanks to my entire family for their support, encouragement, and patience during the many long nights I was writing.

1

Digital Transmission Fundamentals

Before talking about specific technologies, we need to first understand the fundamentals of digital transmission. When we speak about protocols, we talk about parameters, fields, and various values within these fields as if they were plain text. However, what is actually transmitted over wire, through airwaves, and through computer circuitry is nothing more than electrical current (or optical, depending on the medium). Transmission is the lowest form of communications within a network.

1.1 ANALOG TO DIGITAL

All transmissions within a network must be converted from binary code to electrical or optical signals. All information, including the information appended by protocols, must be sent at this lowest form of transmission. In some cases, additional information is appended at this layer to facilitate timing and error control.

1.1.1 ASCII and EBCDIC

Characters must be converted into binary numbers so that they can be converted into electrical current or optical transmission. There are two standards used for representing just plain text, without any formatting. It is important to understand that these standards are not used for the text that is generated by modern-day word processors or desktop publishing systems. When text is word processed, additional information must be provided by the source so that the receiver knows how the text is to look (italics, bold, underlined, specific fonts, etc.). This formatting is not represented in these two standards, but in proprietary formats handled at the upper layers of protocols. Applications receive envelopes of data that include binary information regarding the formatting of the text.

The two standards that deal with plain text are the *American Standard Code for Information Interchange* (ASCII) and *Extended Binary Coded Decimal Information Code* (EBCDIC). ASCII was developed by the American National Standards Institute (ANSI), and EBCDIC was introduced by IBM and is used predominantly by their terminal equipment.

ASCII code is much like Morse code. There are 7-bit codes for each character (both upper- and lowercase), supporting 128 characters. Some of the codes represented do not appear on screens but are used as control characters. For example, EOT is the code for end of text, which is the same as end of transmission. Today, modern protocols provide the necessary control information, and the ASCII characters are now encapsulated within the protocol envelope.

The purpose of the code was for use in terminals, which have no processing capability. A terminal receives a serial bit stream of characters in ASCII code and displays those characters as they are received. Today, ASCII code is still used, but as mentioned above, the characters are now encapsulated within the protocol used to transmit the data. Table 1.1 illustrates the entire ASCII code set. Although numbers are represented in the ASCII code set, there is another method of representing numbers called *Binary Coded Decimal* (BCD).

The EBCDIC code set is almost identical to the ASCII code set, with the exception of an extra bit. EBCDIC supports more graphical characters than does ASCII. However, in today's networking environment, EBCDIC is rarely used. Modern desktop publishing applications have provided a new means for conveying how information is to be displayed. Table 1.2 shows the entire EBCDIC code set.

BCD is used today to represent numbers. Originally, it was a 6-bit code set used to represent alphanumeric characters. Today, it is a 4-bit set used to represent numbers only. You will find it used within many protocols where digits must be represented. Its advantage is in the number of bits to represent a number. The code set only covers digits 0 through 9, but these can be combined to support every number conceivable, with fewer bits than would be required with ASCII or EBCDIC. BCD is used in the Integrated Services Digital Network (ISDN), where telephone numbers must be represented, and in Signaling System #7, a protocol used by telephone companies to convey control information between telephone switches. Table 1.3 shows the entire BCD code set.

One last code set that is used in many telecommunications protocols is the *International Alphabet Number 5* (IA5) code set. This is an International Telecommunication Union (ITU) standard and is very close to the ASCII code set.

TABLE 1.1 ASCII Code Set

				6,5,4	000	001	010	011	100	101	110	111
3	2	1	0	Hex	0	1	2	3	4	5	6	7
0	0	0	0	0	NUL	DLE	SP	0		P		p
0	0	0	1	1	SOH	SBA	!	1	A	Q	a	q
0	0	1	0	2	STX	EUA	"	2	B	R	b	r
0	0	1	1	3	ETX	IC	#	3	C	S	c	s
0	1	0	0	4	EOT	RA	$	4	D	T	d	t
0	1	0	1	5	ENQ	NAK	%	5	E	U	e	u
0	1	1	0	6		SYN	&	6	F	V	f	v
0	1	1	1	7		ETB	'	7	G	W	g	w
1	0	0	0	8	PT	EM	(	8	H	X	h	x
1	0	0	1	9	NL	SUB	)	9	I	Y	i	y
1	0	1	0	A			*	:	J	Z	j	z
1	0	1	1	B		ESC	+	;	K	[	k	
1	1	0	0	C	FF	DUP	,	<	L	\	l	
1	1	0	1	D		SF	-	=	M	]	m	
1	1	1	0	E		FM	.	>	N	^	n	
1	1	1	1	F		ITB	/	?	O	_	o	

TABLE 1.2. EBCDIC Code Set

Column groups are defined by bits **7,6** (00, 01, 10, 11) and, within each group, by bits **5,4** (00, 01, 10, 11). The hex value shown is the low nibble (bits **3,2,1,0**).

Bits 3,2,1,0	Hex	7,6=00 · 0	1	2	3	7,6=01 · 4	5	6	7	7,6=10 · 8	9	A	B	7,6=11 · C	D	E	F
0000	0	NUL	DLE			SP	&	-									0
0001	1	SOH	SBA					/		a	j			A	J		1
0010	2	STX	EUA		SYN					b	k	s		B	K	S	2
0011	3	ETX	IC							c	l	t		C	L	T	3
0100	4									d	m	u		D	M	U	4
0101	5	PT	NL							e	n	v		E	N	V	5
0110	6			ETB						f	o	w		F	O	W	6
0111	7			ESC	EOT					g	p	x		G	P	X	7
1000	8									h	q	y		H	Q	Y	8
1001	9		EM							i	r	z		I	R	Z	9
1010	A					¢	!	¦	:								
1011	B					.	$	,	#								
1100	C	FF	DUP		RA	<	*	%	@								
1101	D		SF	ENQ	NAK	(	)	_	'								
1110	E		FM			+	;	>	=								
1111	F		ITB		SUB	|	¬	?	"								

TABLE 1.3   BCD Code Set

0000	0
0001	1
0010	2
0011	3
0100	4
0101	5
0110	6
0111	7
1000	8
1001	9

1.1.2   Digitizing voice

Digitizing voice is not a complicated problem. The voice is sampled at regular time intervals. Each time the voice is sampled, the amplitude (height) of the signal (waveform) is compared to a scale, which consists of numbers, beginning at 0 and ending at 256. There are both a positive side of the scale (0 to 256) and a negative side of the scale (0 to 256). Figure 1.1 shows the scale and how an analog waveform is sampled. Each time a sample is taken, a pattern is created. The shaded bars in the figure indicate this. The bars represent what is called the *pulse amplitude modulation* (PAM) signal. This signal indicates the amplitude of the signal at the time the sample was taken.

The scale that is used to create the PAM signal is called the *quantizing scale*. This scale is limited in the steps it can repre-

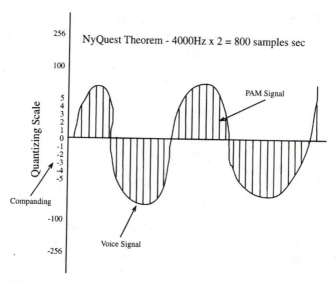

Figure 1.1 Pulse coded modulation (PCM) and pulse amplitude modulation (PAM).

sent, which limits it to the amplitude it can represent. If all of the steps are equally divided on the quantizing scale, the waveform may be misrepresented. The trick is to take samples often enough so that the PAM signal is close to the original analog sine wave. The more often samples are taken (frequency) the better the regenerated signal. For a voice signal, this is 8000 samples per second.

The next step after deriving the PAM signal is to create the digital equivalent of the PAM. The numbers on the scale are converted into 7-bit binary numbers, with the most significant bit (the first bit transmitted) identifying whether the number is a positive or negative value on the scale. The outcome of the digital conversion is called the *pulse coded modulation* (PCM) signal. It is a binary stream of 8-bit words, each byte equaling one PAM sample. There are some issues with PCM.

If samples are not taken often enough, the original waveform will not be digitized accurately. This is referred to as *aliasing*. To prevent aliasing, samples are taken at more frequent intervals. The NyQuest Theorem was written by a scientist (NyQuest) who determined that the normal voice frequency is from 300 to 3200 hertz (Hz; cycles per second). To prevent interference from adjacent facilities, 4 kilohertz (kHz) is allocated for analog voice signals.

This being the case, NyQuest figured that the sampling rate should be twice that of the frequency, which in the case of voice would be 8000 samples per second. Although this theory helps prevent aliasing, electrical properties can still introduce noise that can affect the PAM signal. Low-pass filters built into the electronic circuitry of the devices that provide analog-to-digital conversion also prevent aliasing by smoothing the PAM signal and making it more "readable."

Since the amplitude in most voice conversations is in the midrange of the quantizing scale, it makes sense to put more steps in the midrange and fewer steps in the upper and lower ranges. This provides better representation for "everyday" conversation while still accommodating the infrequent signals at higher or lower amplitudes. This is referred to as *companding*. The tradeoff is distortion of sorts when the amplitude goes beyond these ranges, but the distortion is acceptable for voice communications. If it were used for high-fidelity audio, this would not be acceptable. There are two forms of companding, mu-law and A-law. Mu-law is the method used here in the United States, and A-law is the method used in international networks. The two are very similar but not compatible. There are other types of PCM, which obtain the same results but offer compression. Adaptive differential PCM uses a 4-bit word rather than an 8-bit word. Each 4 bits represent a change in amplitude. This allows two devices to be connected to one port, sending pairs of 4-bit words (one pair for each device).

Another form of compression is called *digital speech interpolation* (DSI). This compression technique is found in many digital recording devices used in telecommunications networks, including voice mail and voice recognition systems. DSI deletes pauses where speech is not present, saving bandwidth and eventually disk space at the host device.

2

The Basics of
Telecommunications Protocols

A *protocol* is best defined as a set of rules. In digital communications, protocols determine where certain information will be
found in the binary bit stream. Addresses, control information,
user data, and various other fields must be clearly defined and
consistent within each transmission. This is the job of the protocol. Not all protocols are alike. Many protocols provide similar features, but all are very different both in format and in
implementation.

2.1 PROTOCOL LAYERS

Protocol functions are divided into *layers*. This layered approach allows for better segregation of protocol functions as
well as software modularity. Layers are important in communications networks because they allow software upgrades to be
deployed without affecting every node in the network. Only
those devices that use functions within the specific layer need

to be upgraded. In a communications network, the first three layers are the most critical. All layers above the first three have no effect on the network itself. They are resident in end node software and are transparent to network devices (such as routers and switches).

Each layer can be independent, transparent to the layers above and below. The only requirement is that the interface used to communicate between the layers must be compatible with the adjacent layers.

The operating system provides the interface between layers, which helps provide some standardization. In the upper layers, a proprietary interface is typically used to support the specific functions of the protocol. The lower layers can be connection-oriented, whereas the upper layers are connectionless (and vice versa). By allowing the network layer to be connectionless, data units can be transmitted over any available route, helping prevent congestion over any one link.

2.1.1 Physical layer (layer 1)

Layer 1 converts the data unit, including the protocol headers, into an electrical or optical signal for transmission over the network. It does not add a significant amount of overhead to a transmission (except in the case of the Synchronous Optical Network, or SONET). Examples of the physical layer include Ethernet, Switched 56, and SONET.

2.1.2 Data link layer (layer 2)

The function of layer 2 is to provide node-to-node communications. The protocols that operate at this layer are not concerned with the contents of the user data or the data that resides at higher layers. In fact, protocols at this layer have no visibility to a data unit's final destination. The data link's only concern is the transmission of data between any two devices within a network. It provides error detection and correction (to a limited degree) and controls node-to-node communications.

2.1.3 Network access layer (layer 3)

The network access layer is responsible for the transfer of data between the host computer and the network. It is the function of this layer to ensure reliable transfer of data from the source, through the network, to the final destination. Addressing for both the source and the destination can be found at this layer.

2.1.4 Transport layer (layer 4)

The transport layer ensures that data is received in the same order in which it was transmitted. This is not a problem if the source and destination are connected directly to one another. In packet-switched networks, this is not the case. Related data units can take multiple paths, arriving at the destination out of order. The transport layer has the task of placing the received data units back into the order in which they were sent. This layer is resident in host software and does not interact with network devices. The originating host software communicates with the destination host by sending protocol messages and negotiating sessions.

2.1.5 Session layer (layer 5)

It is the responsibility of layer 5 to maintain a dialog with the destination host application in a connection-oriented protocol.

In the days of mainframes, the session layer was what started the session (remember logging in?) and then releasing the session when the dialog was through. A dialog is the communications between two entities. This layer is only needed when connection-oriented services are provided.

2.1.6   Presentation layer (layer 6)

The presentation layer is used with terminals to communicate how characters should be displayed on the terminal. Terminals do not have any processing capability, so a protocol is needed to identify how text should appear on the terminal screen (in the past graphics could not be supported on terminals, at least not as we know them today). Encryption and compression are now provided at this layer.

2.1.7   Application layer (layer 7)

Layer 7 provides management functions to support distributed applications. This is also the interface to user applications where network communications are necessary. For example, an e-mail application will need to communicate with the application layer to prepare an e-mail for transmission over the network.

2.2   PROTOCOL ADDRESSING

Addressing in protocols takes many forms. In all protocols, both the physical source and destination devices must be identified by an address (commonly called the *machine address*). The machine address can be hard coded in an interface card or administered through software. When a computer is connected to a local access network (LAN), the network interface card (NIC) provides the machine address for each device.

In addition to the machine address, a logical address used by the various network devices is often used. For example, in Transmission Control Protocol/Internet Protocol (TCP/IP) networks, machine addresses cannot be supported. Instead, a logical address is assigned, which a network server later converts to a machine address. The logical address typically identifies a user rather than an actual machine.

Once a data unit reaches its destination, the host must determine to which application within software the data unit should be sent. This is identified by a *process address.* There are a number of different methods used to identify processes. Operating systems use *sockets,* which are nothing more than logical ports used to connect to applications. If we reexamine addressing, we can see that the machine is assigned an address (used by the second layer), the user is assigned an address (used by the network layer), and the process is assigned an address (used by the operating system to determine how to route the data unit internally). This further illustrates how layering works within various protocols, allowing functions to be addressed using a hierarchical approach.

2.3   PROTOCOL SERVICES

Protocols provide two levels of service, connectionless and connection-oriented. The use of these services depends on the application. Applications that do not require highly reliable transmission, and are not in real time, typically use connectionless services. Other applications that require guaranteed delivery will use connection-oriented services.

2.3.1   Connectionless services

In a connectionless protocol, data units are sent with all of the information necessary to process the data unit when it is received. There is no guarantee that the data unit will actually be received by the destination since these protocols do not use any sequence numbers and do not provide ordered delivery.

In a connectionless protocol, no session is requested, and sequenced delivery is not guaranteed. This makes connectionless protocols less reliable than connection-oriented protocols. When using highly reliable transmission facilities (such as fiber optics), this is not usually an issue.

2.3.2   Connection-oriented services

In a connection-oriented protocol, a request must be sent first. The request is used to ensure that the destination has the necessary resources to process the data unit once it has been received. The protocol will also support sequencing to ensure ordered delivery. Connection-oriented protocols typically establish a session with the application at the destination, and they maintain that session until the data transmission is complete. This may require transmission of many data units, depending on the size of the user data.

It should be noted here that a session is a logical connection, not a physical one. This is controlled by software, with interaction by the operating system. A host may support multiple sessions simultaneously, depending on the operating system and the platform (hardware).

2.4   PROTOCOL TASKS

Protocols perform a number of specific tasks. Not all protocols provide the same tasks, and if they do, they rarely provide them in the same fashion. Protocols are developed to meet a specific need within a network and are often designed with the network topology in mind. For this reason, protocols can be very different from one another.

2.4.1 Segmentation and reassembly

Segmentation allows a protocol to take the user data and divide it into smaller blocks of data before transmitting it over the network. There are many considerations when determining the size of the data block. Different protocols have different requirements and may support different applications.

The transmission medium can also play a factor when determining the maximum size of a protocol packet. Some media may be limited to the amount of data that can be sent reliably in one block. Others may not have any limitation at all.

A factor called *latency* must be considered when designing networks to support real-time applications such as voice and video. If large blocks of data were to be supported, the receiver would have to wait until the entire block of data had been received before processing the data. Once all of the data was received, it could begin processing it, but if there were a large amount of data to process, there could be noticeable delay.

In the case of voice, this would be apparent by pauses in the voice transmission. The receiver would have to wait for the next large block of data to be received before it could begin processing the next transmission. This introduces more delay as the receiver waits to process the large data blocks. Video shares the same issues.

If an error were to occur, and data was retransmitted, large data units would take a long time. This is another reason why protocols favor smaller data units. They do not require large receive buffers, and retransmission does not burden the network.

In some types of networks (such as packet networks), data is not guaranteed to be delivered in the same order it was sent. This is because data units can travel many different paths, depending on the status of the various network elements.

The protocol must be capable of determining the original order in which the data units were sent and whether or not segments are missing from the original transmission, and they must be able to put the data units back together before giving them to the application. This process may take place at different levels. For example, an application may pass a large amount of data down to the next layer, where a protocol is interfaced. This protocol may divide the data units into smaller data units while appending protocol control information (overhead). The protocol may then pass each new data unit to the next layer, which could repeat the process by further dividing the data unit into yet smaller data units, appending its own overhead, and passing it to the next layer. This could take place at four layers before the data is finally sent.

Each layer works independently, the upper layers being transparent. This means the lower layers know nothing about the contents of the data units at the upper layers. They simply pass the data along and never process the information appended by upper-layer protocols. This can add to the complexity of a network when network boundaries must be crossed.

Reassembly is accomplished through a variety of techniques. A protocol will usually assign a number or identity to each data segment, which, when received at the destination, identifies when the data was sent (by order). Another method is to identify where in the original block of data the data segment belongs. Providing an offset, which identifies to which byte the segment belongs, does this. For example, if a data segment starts at the thirteenth byte of data, the offset would be byte 13. The receiving host knows when reassembling data to place this segment after the twelfth byte.

2.4.2   Encapsulation

All protocols perform some form of encapsulation. This is the placing of data into an "envelope" of sorts, surrounded by protocol control information. An example of control information may be source and destination addresses, as well as error-checking data. This forms a *packet,* frame, or protocol data unit, depending on where the data is being processed within the protocol stack.

As the data is passed from one protocol layer to another, additional protocol information can be appended around the existing data, causing the data unit to grow as it passes through each layer. Figure 2.1 illustrates how user data is passed to a file transfer protocol, which adds control information and then passes the data unit to the next layer.

The protocol at the next layer appends information necessary for the host to maintain error control and addressing so that the receiving host can route the data to the appropriate application. This information is then passed to the next layer, which encapsulates

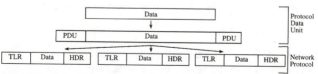

Figure 2.1   Protocol flow with encapsulation and segmentation.

the original data as well as all of the appended information into its own envelope.

This next layer may add addressing for routing through a network. In addition, error control information is needed as well as sequence numbering (for reliable data transfer). This information is then passed to yet another layer, and so on. As one can see, as data is passed through the various layers and finally makes its way onto the network, the size of the data unit grows.

2.4.3   Connection control

Connection control is found in connection-oriented protocols only. It is the responsibility of the protocol to first establish a connection with the destination. This is not a physical connection, but a logical one. This is accomplished by sending a variety of protocol messages to the destination host and waiting for acknowledgments.

Once these messages have been acknowledged, user data can begin flowing through the network. As the data is received, the destination sends periodic acknowledgments to notify the originating host that the data has been received.

When data transfer is complete, it is up to the protocol to notify the destination of connection termination. This means the logical connection is released, and the resources being used by the destination host can now be used for another session. In circuit-switched networks, where physical connections are established, this means the release of the physical link between devices.

2.4.4   Ordered delivery

Ordered delivery is accomplished by numbering each data unit as it is passed to the network. The receiving host then keeps track of each of the sequence numbers as it is received. When sending acknowledgments, the receiving host identifies the sequence numbers it has received.

Most protocols do not require an acknowledgment every time a data unit is received. Instead, the receiving host waits until several data units have been received. It then sends one acknowledgment for all data units received. This means that both the receiving and originating hosts must maintain buffers.

A transmission buffer keeps all data units that have been transmitted until they are acknowledged. Once they have been acknowledged, they can be dropped from the buffer. If a retransmission is requested, the originating host retransmits everything in its transmission buffer.

A receive buffer is used to store all data units until they can be processed. As resources become too busy to handle the received data units, congestion occurs, and buffer overflow causes errors. This buffer is also used to collect associated data units, which are those that have been segmented and must be reassembled by the receiving host.

2.4.5 Flow control

Flow control is important at all layers of the protocol stack. Think of a printer. As the memory (receive buffer) becomes full, the printer must be able to notify the sending host that congestion has occurred and that it should wait before sending more data. The same is true within a network. Protocols must be able to control the flow of data through the network to prevent errors.

Some protocols use the sequence numbers for flow control. Others add an additional parameter that indicates how many data units can be received before an acknowledgment is required. In all cases, special messages can be used to stop the flow of data units if the receiving host is no longer capable of processing data units.

2.4.6   Error detection and correction

Error detection and correction is another process provided by protocols. Sequence numbers are added to the header of a packet and are used to ensure ordered delivery. This allows the receiver to determine whether the data units have all been received and if they have been received in the same order in which they were sent.

Part of the header also contains information used to check the integrity of the data received. There are several methods used for checking integrity, but they all use the same basics. An algorithm is run before the data is actually transmitted, and the results are placed in the header. When the data has been sent, the same algorithm is run again, and the results checked against the value placed in the header.

In both cases, when an error is detected by the protocol, it cannot be fixed. Instead, the protocol discards the data unit and returns an error message to the sender. Depending on the protocol, this message is treated as a request for retransmission. In some cases, the actual message is called a retransmission request, whereas in others the message is simply an error message.

2.5   DIGITAL TRANSMISSION

To transmit more than one conversation on a single copper pair, the voice transmission must first be converted to digital. We have already discussed the process used to convert voice transmission to digital form. The PCM transmission is inserted into one of several channels transmitted over the cable pair and is dedicated to this channel for the duration of the call.

Once we have converted voice to digital form, any medium can be used to transport the transmission to its destination. Fiber optics has proven to be the best alternative for a number of reasons. The cabling itself is flexible and is not subject to external interference caused by electrical equipment. Copper, on the other hand, is very susceptible to external interference.

2.5.1 Multiplexing

To send multiple transmissions over the same copper pair, the transmissions must be divided. This is referred to as *multiplexing*. There are two principal forms of multiplexing: frequency division multiplexing (FDM) and time division multiplexing (TDM). Both of these techniques achieve the same result: multiple transmission over the same medium.

FDM divides the bandwidth of an analog medium into blocks of frequencies. If you remember our discussion about digitizing voice, you will remember that voice requires 4 kHz of bandwidth. In FDM, the transmissions are divided into 4-kHz frequency blocks, starting with 0 to 4 kHz. The next block begins at 4.1 kHz and ends at 8 kHz, and so on. Table 2.1 shows how FDM is divided into blocks and the total number of transmissions supported by FDM.

Transmissions are sent simultaneously, but the multiplexers change the actual frequency of the voice transmission according to this hierarchy. This means that a demultiplexer at the terminating end must convert the transmission back into its original frequency. A 12-channel multiplexer will output 48 kHz of bandwidth, divided into twelve 4-kHz blocks of

TABLE 2.1 Frequency division multiplexing hierarchy.

Mux. Level	No. of Voice Circuits	Formation	Frequency Band (kHz)
Voice channel	1		0–4
Group	12	12 voice circuits	60–108
Supergroup	60	5 groups	312–552
Mastergroup	600	10 supergroups	564–3084
Mastergroup mux.	1200–3600	various	312/564–17,548
Jumbogroup	3,600	6 mastergroups	564–17,548
Jumbogroup mux.	10,800	3 jumbogroups	3,000–60,000

transmission. Each block is considered a channel, supporting one transmission.

FDM is more expensive than TDM, which is a digital technique. There are other problems with FDM. Because it is analog, it is susceptible to external interference. Noise (in the form of static and "pops") is simply amplified and cannot be effectively filtered out of the transmission. There are very few FDM facilities used today.

2.5.1.1 Time division multiplexing. The most common form of multiplexing is TDM. This digital form of multiplexing is used in virtually every digital network today. The transmission facility is divided into time slots. Each time slot is used to carry one transmission.

TDM was developed for digital transmission facilities such as T-1. TDM uses the same concepts as FDM, but instead of using blocks of frequencies, it uses time slots of 64-kbps increments. Digital transmission is linear in fashion, meaning that bits are sent in serial order.

Each transmission is assigned a time cycle, or time slot. When this time slot occurs, data from the assigned device is transmitted. Think of the gates used in horse racing. All of the horses are lined up in the gate. In TDM, only one gate opens at a time, allowing data in the queue to be transmitted.

Once the gate opens and the queued data is sent, the gate is closed and the queue is filled with new data. When the time slot comes around again, the gate is opened and the data in the queue is transmitted.

Our discussion of PCM and voice digitization said that digital voice (PCM) is sent in 8-bit segments. If we take that 8-bit PCM "word" and place it into a digital time slot, we can support up to 24 individual time slots, each carrying one 8-bit PCM word from a different transmission.

The time slots are repeated; that is, channel (or time slot) 1 is repeated every 24 channels. A telephone call sent over a digital facility is assigned to a dedicated channel for the duration of the call, which means the conversation will be converted to digital form and transmitted in the same channel until the call is terminated.

A 24-channel block is called a *frame*. Table 2.2 shows that one frame is called a digital signal level 1 (DS1) in the United States. Time slots, or *channels* as they are also called, are fixed in length and can be used to send voice or data.

2.5.1.2 Synchronous and asynchronous TDM. There are two methods of TDM: asynchronous and synchronous. In synchronous TDM, the time slots are assigned to a device. If the device has nothing to send, the time slot passes with no data. It is considered "idle." This is not an efficient use of bandwidth because no device is transmitting 100 percent of the time, and many time slots will pass with nothing in them.

In asynchronous TDM, time slots are dynamically assigned. If a device has something to transmit, it is assigned the next available time slot. If there is nothing to transmit, the time slot goes by with nothing sent (unless another device has something to transmit, in which case the time slot can be assigned to that device).

Asynchronous time division multiplexing (ATDM) is the concept used for newer technologies such as asynchronous transfer mode (ATM). The transmission facility uses bandwidth efficiently, with every time slot being used. Only when no device has anything to send are time slots empty (idle).

Although ATM is based on ATDM, ATM is not channelized. In other words, there are no time slots in ATM. Data is inserted

TABLE 2.2 North American digital signal hierarchy.

Digital Signal Designation	Bandwidth	Channels (DS0s)	Carrier Designation	Medium (typical)
DS0	64 kbps	1 time slot	None	Copper, fiber
DS1	1.544 Mbps	24 channels	T-1	Copper, fiber
DS1C	3.152 Mbps	48 channels	T-1c	Copper, fiber
DS2	6.312 Mbps	96 channels	T-2	Micro, fiber
DS3	44.736 Mbps	672 channels	T-3	Micro, fiber
DS4	274.176 Mbps	4032 channels	T-4	Micro, fiber

in a fixed-size cell, and the cell is then transmitted. Parameters within the cell (header) provide the necessary information for routing and processing the data.

2.5.1.3   Statistical ATDM.   Another form of ATDM is statistical ATDM, which requires some intelligence in the multiplexing equipment. When a device has data to transmit, how much data must be sent and how many time slots will be needed for that transmission are determined. The multiplexer then reserves these time slots for the transmission. The result is several consecutive time slots assigned to the device, rather than the device being assigned one time slot at a time and waiting for another idle time slot to become available. Devices with less data are allowed to transmit in between the time slots assigned to the "chatty" device, preventing devices from becoming congested with data waiting to be sent.

In ATM, another parameter has been added. Referred to as *quality of service* (QoS), this parameter indicates whether or not data in the queue waiting to be transmitted is delay-sensitive data. Voice and video are considered delay-sensitive since delaying their transmission will affect the quality of the received transmission (in other words, these are real-time applications, and delays will cause pauses in the playback of the transmissions). QoS works like a priority, indicating that a device may have delay-sensitive data that should be sent ahead of any other data waiting in queues.

2.6   THE U.S. DIGITAL HIERARCHY—DS1 AND DS3

The digital hierarchy used in the United States, which is shown in Table 2.2, starts with the digital signal 0 (DS0).

This is the lowest common denominator in the digital hierarchy. Each DS0 is used to carry a transmission, usually voice or data. As shown in the table, 24 DS0s make a DS1. A DS0 supports 64 kilobits per second (kbps). In a DS1, which uses 24 DS0s, 1.544 megabits per second (Mbps) is supported.

A DS1 has specific bits that identify where the first and the last DS0s are located. The entire DS1 is referred to as a *frame.* Protocols such as T-1 provide the framing bits as well as some other signaling information required delineating the 24 separate transmissions.

A device such as a channel bank or digital cross connect allows the DS1 to be terminated in a central office or subscriber premise, and it separates the various DS0s from the frame. Once separated, they can then be connected to various pieces of equipment, such as a telephone switch or router. Some devices have the capability of connecting directly to a DS1 without the use of a channel bank or cross connect by providing the same demultiplexing function within the device.

A DS3 is made of 28 DS1s, which must be demultiplexed before connecting to a device. Again, a channel bank or cross connect is needed to perform this function. All of these devices use TDM, taking the digital voice or data and placing it into time slots. When multiplexed, the individual DS0 time slots are placed into a frame and transmitted to the distant end, where the DS1 is demultiplexed and the individual DS0s are extracted from the frame and routed to the network equipment.

The digital hierarchy was designed to consolidate facilities between central offices. Instead of using a pair of wires for only one transmission, multiplexing allows telephone companies to send many transmissions over the same pair of wires.

2.6.1   European digital hierarchy

It should be mentioned here that Europe also uses digital trunking. However, the digital hierarchy used throughout Europe and the rest of the world does not match that used within the United States. Theirs consists of 32 channels, providing more bandwidth within one frame. The European hierarchy also uses 64-kbps building blocks (see Table 2.3).

2.6.2   T-1 facilities

T-1 facilities use the DS1 carrier as a transport. The T-1 protocol adds necessary overhead to the DS1 for framing and loopback testing, among other things. When extended to a subscriber premise, T-1 is used to provide an interface between the subscriber and the telephone company. Other protocols can be used over the T-1 because the T-1 does not attempt to process any information other than the specific bits used for framing and signaling at the physical layer.

T-1 circuits use what is referred to as a *D-type* channel bank. The channel bank provides the multiplexing/demultiplexing functions required for digital carrier systems. The T-1 uses specific bits within the digital transmission for signaling and synchronization of the frames.

A typical T-1 circuit consists of at least four wires, two for transmission and two for receiving. Remember from our discussion that voice when digitized is sent in PCM format. A PCM word consists of 8 bits for each sample of the voice. This 8-bit word is then assigned to a channel, or time slot, within the T-1.

Figure 2.2 shows how bits from the transmission are "stolen" and used for framing and signaling. There is no degradation of transmission from this bit robbing, at least none

TABLE 2.3 CEPT digital signal hierarchy.

Digital Signal Designation	Bandwidth	Channels (DS0s)	Carrier Designation	Medium (typical)
Signal Level 0	64 kbps	1 time slot	None	Copper, fiber
Signal Level 1	2.048 Mbps	30 time slots	E-1	Copper, fiber
Signal Level 2	8.448 Mbps	120 time slots	E-2	Copper, fiber
Signal Level 3	34.368 Mbps	480 time slots	E-3	Micro, fiber
Signal Level 4	139.264 Mbps	1920 time slots	E-4	Micro, fiber
Signal Level 5	565.148 Mbps	4032 time slots	E-5	Micro, fiber

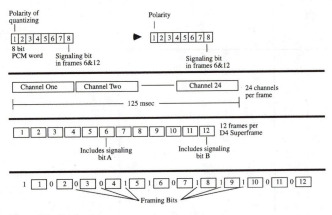

Figure 2.2 D4 framing.

discernible by the human ear. This is because only 1 bit out of an 8-bit word is taken, representing only one sample of the analog voice. The eighth bits from the sixth and twelfth channels are used for signaling in the D2, D3, and D4 channel banks.

Twelve frames of DS1 are referred to as a *superframe*. The type of channel bank used at both ends of the circuit determines where the framing bits are located within a frame. It is important for a T-1 to use the same type of channel bank at both ends of the circuit.

The Bell System has defined specifications for channel banks, specifying which bits are used for signaling and which are used for framing. For example, in the first generation of channel banks, D1, bit 8 of every channel was used for signaling (on hook, off hook, etc.). However, in later generations of channel banks (D2, D3, and D4) every sixth and twelfth frame of a superframe (defined below) contains the signaling bits. The eighth bit in these channels is used for signaling. There are 192 bits in a frame, plus 1 extra bit used for synchronization, for a total of 193 bits in a frame.

An extended superframe (ESF) is a combination of 24 frames, each frame being 24 DS0 channels. In other words, 24 DS1s become an extended superframe. Framing bits are inserted in frames 18 and 24. There is no difference in the way the DS0s are used; it is simply a matter of being able to combine many more DS0s into a collective group. This should not be confused with the other levels of the digital hierarchy. ESF is simply an additional multiplexing function for the transmission of DS1s. If additional bandwidth is required, a DS3 or higher is used to support applications (such as video, which uses DS4 and DS5).

2.6.3 Synchronization

Synchronization is critical in any digital carrier system. If a device is connected at the far end of a circuit, and it is not synchronized with the near end channel bank, the far end device will not know when a T-1 frame begins or ends. It will misinterpret the boundaries of the frames, resulting in transmission errors. Synchronization is achieved through the use of highly accurate clocks. These clocks are then synchronized to the same reference, using satellite to access the reference signal. Clocking is critical in any digital trunking network.

3

Computer Telephony Applications

A new industry has been created that uses computers in telephone networks. Users of private telephone systems have tried for many years to marry the computer with the telephone system. Until the last 5 years, the industry had been unable to meet the needs of the user in this area.

Several manufacturers of private branch exchange (PBX) and computer equipment finally got together and forged agreements to develop interfaces that would allow desktop computers to control applications in PBX equipment. Some of the first applications were things like voice mail systems and calling center applications. Today, one can even buy plug-in computer boards that allow your computer to act as a mini-telephone system.

The industry is now moving toward using TCP/IP data networks for the transmission of voice communications. These interfaces are now being adapted to work with TCP/IP and a plethora of new devices providing call control and transmission.

This will further the use of desktop computers for applications such as video conferencing and voice transmission in enterprise networks.

Today, you can change the routing in your PBX, change voice mail configurations, and even receive sophisticated reports for your calling center, all through a desktop computer. Although there is now a standard, not everyone has endorsed this standard. In fact, the European vendors were the first to pursue and provide a standard interface to their PBX systems for computer applications.

3.1   Telephony Application Programming Interface (TAPI)

The European Computer Manufacturers Association (ECMA) started developing the Telephony Application Programming Interface (TAPI) in 1988. Understanding the complexities of creating such a standard, and having difficulty focusing on any one solution, the ECMA formed a technical group (TG11) that is responsible for defining the standard.

The ECMA drafted TR52, which is not a standard but rather a report citing the various solutions found most favorable by TG11. This approach was taken because the group found many different approaches to solving the same problems, and it was becoming difficult to agree on any one solution. To expedite the standards process, this report provided important insight to each solution.

The workgroup's first task was to outline the specific applications to be targeted so that everyone would have a clear understanding of what they were trying to accomplish. Efforts up to this point had led to many different proprietary solutions for the same applications. The applications targeted by TG11 were

- Call centers (inbound and outbound)

- Customer support environments

- Emergency call applications (such as 911)

- Data collection and distribution

- Data access

- Hotel applications

- Switched data

Call centers are commonly found at airline and hotel reservation centers. They typically consist of many operators, or "agents," who have specialized telephone sets. Every call handled by an agent is tracked by computer reports, allowing the call center manager to determine whether or not there are enough agents on duty to handle the flow of traffic. These reports can also alert the call center manager to traffic patterns, which are then used to determine when additional 800 numbers need to be assigned to the group. Even trunk utilization can be determined by the reports offered by many systems.

Productivity reports allow call center managers to monitor the group's activities and compare agents with other agents, groups with other groups, or even agents with entire groups. All of these reports, and in many cases the actual configuration of the call center (number of trunks, number of agents, etc.), are accessible through a desktop computer, which in turn communicates with the PBX through the TAPI interface.

For outbound calling centers, a database selects telephone numbers randomly (usually based on some demographic or geographic parameter selected by the calling center management).

Before the computer dials the call, the next available agent is determined. The computer server then activates a screen on the agent's computer, showing the name, address, and telephone number that the computer is about to call. The agent can allow the call to proceed or interrupt the call and go to the next selection.

In a customer service environment, the desktop computer can be linked to network management software. As a computer user calls the help desk to get assistance with a computer application, or possibly to report network connection trouble, the help desk agent can be looking at the user's network connection.

The interface in this type of environment can provide information regarding the caller's extension number, which when linked to a database, can automatically trigger the network management software to identify the network address of the caller's computer. From there, the help desk can quickly troubleshoot the network connection. This saves the agent the time of looking up the caller's network specifics before troubleshooting.

The computer used by the help desk can even link calls to contact management applications so that the customer service representative can keep a history of the telephone call. If a database for customers is available, this can also be activated with the caller's data on the screen, making it simple for the customer service representative to handle the call.

We are all familiar with emergency call applications such as the 911 system used in most U.S. cities today. Another application for this can be found in some condominium complexes and hotel systems. When someone dials 911 from a PBX system, the 911 system cannot identify the extension number of the caller, and the only information provided is the main billing number of the business owning the telephone system. This can be a perplexing problem in a large hotel, making it almost impossible to identify where the distress call originated. Not only that, but legislation passed regarding 911 support makes it illegal. The PBX industry has been able to ignore this issue for years, partly because the focus has been on the wireless industry; however, the focus is now back on the PBX industry, and manufacturers are being forced to provide resolutions to the 911 problem.

At least one manufacturer addressed this through the use of a computer. The computer is attached to a TAPI interface. Whenever someone dials 911 from a PBX extension, the PBX sends the extension number of the caller to the computer. The computer searches a customer-defined database for the extension and finds the room number of the caller. Attached to the computer is a digital display, which is used to display the actual room number of the caller who dialed 911. This works equally as well in a condominium complex where a PBX is used to re-sell dial tone to the condo owners. The display unit is placed in a conspicuous area where emergency workers are most likely to see it when they arrive.

Data collection and distribution was one of the earliest applications for computers in telephone systems. For example, every PBX sends out calling data in the form of detailed reports identifying every telephone call made. These reports provide the extension number that placed the call, the duration of the call, the time the call was placed, and the number that was dialed.

This information is difficult to use in its raw format and would require someone to collate and organize each report since there is no sorting in the raw report. This is where the computer comes in. The computer is attached to the PBX and receives the detailed report at certain intervals or in real time. The computer then sorts the data and provides a call activity report that can be sorted by extension number, department number, or even user name.

These reports can even have costs assigned to various types of calls for cross-charging departments or charging guests (as is the case in hotel systems). This is only one example of data collection and distribution in a telephone system.

In calling centers, agents need access to databases that provide them with information about the caller (or in the case of outbound call centers, information about the called party). These databases can provide important information, including the customer's address, billing information, and even purchasing profile.

Some of these call centers have even automated the task of placing orders by asking the caller to enter a customer number or his or her own telephone number for routing to an automated voice response system. Once into the voice response system, the customer uses a dialpad to enter ordering information, such as product number. The database provides the rest of the information needed to process the order. The billing information and shipping address are already known.

Hotels have many applications that require computers. We have already mentioned the call accounting system, used to track and record telephone calls made by guests. These call accounting systems automatically mark up the calls and send the information to a room billing system to be added to the room bill automatically.

Hotels also use automated directories. These directories allow operators to connect calls to guests without having to look up the guest's name on a sheet. Every guest is entered into the computer upon check-in. The computer creates a directory, which the operator can then search by entering in the guest's name. Once found, the call can be transferred to the room by simply pressing Enter or a similar key on the computer keyboard.

Switched data was not addressed by the ECMA because there was not a lot of interest in this area. A few switch vendors did provide this functionality in their PBXs, however, allowing computers and peripherals to be attached to the PBX rather than a separate LAN. The problems with using a PBX to switch data are capacity and speed. PBXs can only transfer data at 19,200 kbps. This makes it difficult to compete with LANs running at 10 Mbps and up.

Once the applications were defined, a model was established. The model had to define the boundaries in which the standard would operate. This was important for the developers to understand, because otherwise they would spend far too much time trying to understand the operating parameters of their standard. By defining the domains in which the various aspects of the standard would apply, the focus of the various committees could remain within their area of domain. Three domains were identified: computing, switching, and applications. The various committees worked within specific domains and identified the interfaces between the various domains.

Within each domain, objects were identified. These objects would require some form of identification so that the software could maintain the status of that object and be able to correlate the object with a specific call. For example, in the switching domain, a connection identifier acts as a reference for endpoints outside of the switching domain to specify a certain call. A connection is considered a transaction between an endpoint (a device) and a call. The state of the connection is reported to the computing domain so that the computing domain knows what is happening with any one connection.

A device can be a telephone, a button on a telephone, an operator, an automatic call distribution (ACD) hunt group pilot number, a trunk, or any other physical (or logical) entity that is involved in a call. These are all treated as objects within their own domains.

A call represents communications between one or more devices. In any typical situation, a call could be associated with several connections, all which report their connection states (busy, alerting, connecting). In addition to being associated with connections, a call would also be associated with devices. The TAPI standard defines how this is reported from one domain to the other and how the references are managed.

In a nutshell, TAPI is a standard that defines a set of control messages understood by the telephone system and managed by the computer attached to it. Applications that interface to these control interfaces can then be written, allowing programmers to write applications with interfaces to the PBX functions.

3.2 ASAI/SCAI

These protocols were developed for ISDN switches to allow them to send certain parameters passed from the ISDN interface down to databases and adjunct processors. For example, when a call is connected to a PBX over an ISDN interface, the calling party number is passed from the PBX to a database over the ASAI or SCAI interface. Other parameters relating to the origin of the call can be passed over this interface as well. The most common application of this interface is for call centers, where the calling party information is passed over a data communications network to the desk of a customer service representative or reservation agent.

ASAI is an AT&T specification, and SCAI is an IBM specification. Both are similar in operation, providing slightly different parameters and protocol structures.

4

LAN/WAN Networks

A LAN differs from the mainframe network in that there are no centralized processors. All processing is performed on stand-alone computers and workstations. The LAN interconnects these computers, allowing them to share files and communicate with one another.

All applications used by computer users are located on individual personal computers and workstations. This means that if the LAN should fail, the computers are still operational; they just cannot communicate with one another. The popularity of personal computers (PCs) led to the development of LANs, and today, the LAN industry has far outgrown the mainframe industry.

One disadvantage of a LAN is the cost. Although the network costs are relatively low, the cost of individual computers can be rather high. In the mainframe environment, users communicate via terminals, which cost a few hundred dollars. In the LAN environment, users communicate through PCs and high-powered workstations, some costing over $10,000 (depending on the applications for which they are used). Even a simple computer with relatively low processing power can cost a few thousand dollars.

Another disadvantage is the support cost of PCs. Applications continuously change, requiring upgrades. Often these upgrades require additional memory, forcing users to purchase additional memory chips for their computers. The cost of memory has not come down significantly, often costing as much as a new computer itself. Mass storage is fairly cheap, but memory is still rather expensive.

Still, when compared to the ongoing cost of mainframe equipment and productivity loss, the cost of LANs is really not much more than mainframes. In today's world, we have become so dependent on having computers that without one, we would be lost: no word processing, no spread sheets, no checkbook programs. This implies a shift in the computer industry from legacy systems to more personalized systems tailored to meet the needs of each individual user rather than a corporation.

Even this paradigm is changing. Many companies are now offering network personal computers (NPCs) specifically designed for applications that require central storage of information. The NPC is a computer with memory and a processor. However, there are no disk drives in an NPC. This allows the operating system to be significantly smaller. All applications must be run from a network server, reducing the cost of software upgrades. All files are stored on the server as well. This may introduce a bit of a problem in terms of throughput at the server if large workgroups are accessing it at the same time. To alleviate the bottleneck at the server, workgroup servers would probably be used. All files and applications would be stored on the workgroup server, freeing the main server for other functions.

4.1 TOPOLOGIES AND BASIC ARCHITECTURE

When connecting computers together, there are two forms of communications: point-to-point and multipoint. In point-to-point, two PCs are directly connected together and communicate with one another through this interface. In a LAN configuration, more than two computers are connected to this interface, but two computers are capable of addressing one another exclusively.

In point-to-multipoint connections, one computer is capable of connecting to many other computers at the same time. This is sometimes referred to as *broadcast* mode. The computer sends the same message to all the computers in the connection. This connection does not necessarily imply a physical connection. A logical connection can be obtained by addressing specific computer addresses in any mode, point-to-point or point-to-multipoint.

A topology is the logical layout of the network. It may not indicate actual physical connections since many devices can be used to obtain the topology needed without physical wiring. For example, one may want a ring topology, which would require wiring all computers in a daisy chain from one to the next. The same can also be obtained by using a hub device where the ring topology is obtained through internal wiring in the hub itself. All of the computers are connected directly to the hub.

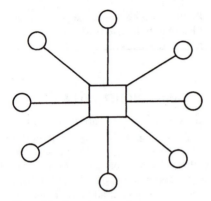

Figure 4.1   Star network topology.

There are several different topologies used in LAN configurations. The star topology uses a hub device, with all devices connecting to the hub (see Fig. 4.1). The hub acts as a switch, allowing traffic from one computer to be switched to the addressed computer. The telephone network uses a star topology. This is not a very popular topology for LANs because the hub can introduce a bottleneck to data throughput.

Another more popular topology is the bus (see Fig. 4.2). In a bus topology, PCs are connected to one common bus. Transmissions from the PCs are placed on the bus for all other PCs to examine. Addressing is used to identify the destination of each data packet. As data is transmitted over the bus, each PC looks at the data packet header to determine if the address is its own. If it is, the PC copies the data into memory. The original transmission stays on the bus, however, and must be absorbed at the end of the bus. Data is transmitted in the form of electrical sig-

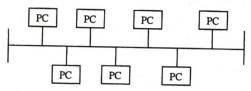

Figure 4.2 Bus network topology.

nals, which means the transmission is capable of reaching the end of the bus and reflecting back toward the originator. Therefore, the bus topology requires a terminator at each end, which acts like a sponge to absorb the electrical signal and prevent it from reflecting back onto the network. Although there are many different bus solutions, probably the most popular is Ethernet. Ethernet is a protocol designed specifically for LANs using a bus topology.

A ring topology interconnects many computers together by daisy chaining one computer to the next. This means that every computer in the ring will have at least two connections: one for data coming in from its neighboring computer and one for data being sent to the next computer in the ring.

In this configuration, data is sent from one computer over the ring to the next computer on the ring. The computer then examines the header of the packet to determine if the address is its own. If not, the data is then retransmitted over the ring to the next computer. This process is repeated until the data packet reaches the final destination.

The fact that every computer in the ring retransmits the data implies that there is a built-in repeater function when using ring topologies. This means that external repeaters are not needed; the electrical signal is regenerated every time the data packet is

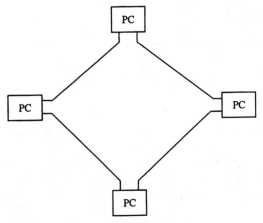

Figure 4.3   Ring network topology.

sent from one computer to the next. This is an inherent advantage when using ring topologies (see Fig. 4.3).

The downside to ring topologies is that every computer must process the data packet as it makes its way through the network. Each computer introduces a delay in transmission while it determines the address and retransmits the data packet. This is different from bus topologies where the data is sent over the bus, and all computers examine the data without having to process the data.

Some ring networks use a dual ring topology. Fiber Distributed Data Interface (FDDI) is one example of a LAN solution that uses a dual ring. The idea is that one ring can be transmitting data in one direction, while the other ring is sending data in the opposite direction. If one ring breaks, the network is capable of healing itself by providing a connection between the two rings and forming one contiguous ring.

As shown in Fig. 4.4, in the dual ring configuration, each node on the ring has both an input and an output for each ring. Traffic is transmitted over both rings, but in opposite directions. By transmitting traffic in opposite directions, any node can become a bridge between the two rings if there is a break.

Another feature of ring topologies is the ability to detect where a ring is broken. This is handled in a number of different ways, depending on which protocol is used. The protocol provides ring management, which usually entails nodes sending

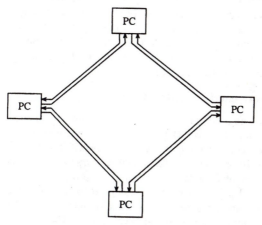

Figure 4.4 Dual ring topology.

management messages to the neighboring node, constantly checking the status of their neighbor.

Dual ring topologies are also used in SONET, providing telephone companies with a more robust alternative to their existing digital transmission facilities. In the existing network, if there is a cable break, the central offices must determine which part of the network is out of service and route traffic around an entire office. With SONET, the ring is monitored by the protocol, and in the event of a break in the ring, traffic is automatically routed to the secondary ring.

4.2 LAN/WAN DEVICES

There are really only three basic network components in a LAN that are used to route data through the various parts of the network. Each one serves a specific function in the network. These devices follow a hierarchical approach in supporting the network. Other devices are used in data networks for connectivity. They use different portions of data protocols depending on their function within the network.

4.2.1 Network interface card (NIC)

Other devices are used in networks besides the three main components mentioned above. The NIC is an integral part of any network. For a computer to connect to the network, it must have an NIC. The NIC provides the functionality of transmitting data over the copper or fiber facility. The NIC is protocol-specific and also provides the machine address used to identify each node in the network. The machine address is "burned" into the NIC by the manufacturer, so it can never change. This ensures that each computer has its own unique machine address.

Computers within a single network cannot have the same machine address; however, computers in different networks can have identical machine addresses (because there is also a network address to separate the two nodes). The general rule of thumb in any network is that every node within the same network must have a unique network "name," which serves as its address.

4.2.2 Media access unit (MAU)

In Ethernet networks, a media access unit (MAU) is used to connect the node to the network and it is a connection point between the LAN and the computer.

The MAU serves several purposes. It determines when a node is able to transmit over the bus, provides a standard interface between the node and the network, and determines if there has been a collision after transmitting data.

Not all Ethernet networks use a MAU to connect the computer to the network. A MAU is typically used with thick coaxial cable. This is commonly referred to as 10BASE5. Networks using thin coaxial cable have NICs with built-in MAUs.

4.2.3 Repeaters

Data is transmitted through a LAN over copper wire. Copper is capable of transmitting data over a limited distance, depending on the type of wire and the protocol used in the network. Protocols can enhance a network's ability to send data over longer distances by changing the representation of the data and by using different compression techniques. However, there is still a limit to how far data can be transmitted before the electrical signal used to represent the data begins to fade. This is when a repeater can be beneficial. Networks using bus topologies are frequent users of repeaters because the bus can only be so long before the original signal must be regenerated. It is important to understand here that repeaters regenerate signals; they do not amplify them.

To regenerate a data signal, the repeater must be able to determine what the original signal looked like. As data is received, the electrical signal is no longer as clean as it was. The waveform becomes more rounded rather than square. The repeater is capable of determining what the original waveform looked like, and it regenerates the signal based on an algorithm it performs.

Not all networks need repeaters. In ring topologies, every node in the network provides the repeater function. As the signal is sent from one node to the next, each node regenerates it. However, if there is a long distance between any two nodes, a repeater may become necessary.

Repeaters are considered layer 1 devices; they do not interpret any part of the data and are not capable of examining any of the addresses in the protocol header. They receive the data signal in the form of an electrical signal (or optical signal) and regenerate the entire signal based on the results of an algorithm.

4.2.4   Bridge

A network bridge is used to join two different network segments (see Fig. 4.5). Understand that a bridge does not join two different networks together. Large networks must be divided into smaller segments to prevent too many nodes from creating congestion. The bridge is then used to connect the segments together.

When we begin discussing protocols, you will find that three addresses are used in a protocol. The machine (node) ad-

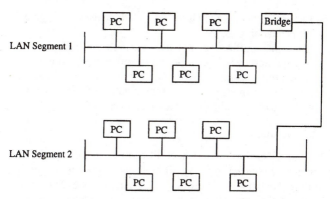

Figure 4.5 LAN bridge.

dress identifies the computer that generated data or is to receive a message. The network address identifies the network that sent data or is to receive the data. The third address type is used to determine which application in the computer sent or is to receive the data; this will be discussed later on.

The bridge is not capable of examining network addresses, which are found in the protocol header. LAN protocols do not use network addressing because they only send data from one node to the next on the same network. Wide area network (WAN) protocols provide interconnectivity between LANs and use network addresses. The bridge is not used in WANs.

When data is sent over a network, all devices receive the data. Each node examines the machine address to determine if the data is addressed to it or not. The bridge also examines the machine address in the LAN protocol header to determine if the machine address is located on the other side of the bridge. If it is, the bridge copies the signal and sends it over the segment that it is attaching to.

The other traffic does not pass through the bridge and is simply ignored by it. The nodes on the other side of the bridge never see this data traffic. In today's complicated data industry, bridges often provide additional functionality, making it hard to draw a line between bridges, routers, and repeaters.

4.2.5 Routers

A router is used to interconnect different networks together. A router may connect to many different networks, switching incoming data from one network out to another one. This is accomplished by providing multiple ports on the router. Each port is capable of receiving and transmitting data.

Routers depend on layer 3 (WAN) protocols. They too are protocol-dependent because they must examine the protocol header of the network protocol to determine the address of the destination network. They do not look at the machine address, which is found in another part of the data packet (and is part of another protocol).

Routers come in many different flavors, supporting multiple protocols and multiple transmission facilities (transmission facilities here refer to T-1, Frame Relay, and other technologies used to interconnect WAN segments).

Routers provide additional intelligence to the network and are often capable of making routing decisions based on the amount of traffic in a particular segment of the network and of routing around failed parts of the network. There are protocols designed specifically for routers, providing information to the router such as address tables and network status.

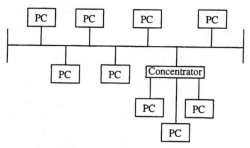

Figure 4.6 LAN concentrator.

4.2.6 Concentrators

A concentrator is often used to make an installation easier (see Fig. 4.6). For example, when installing a ring topology, it may become difficult to run wire from one computer to the next, maintaining a ring. If a concentrator is used, all computers can be wired directly to the concentrator, and the internal wiring of the concentrator forms the ring topology.

4.2.7 Gateways

When connecting to a network using a protocol that is different from that used by others, protocol conversion must be provided. For example, in connecting an Ethernet network to a Token Ring network, a device must be used to convert the protocol headers from Ethernet to Token Ring. This is so the devices in the network can interpret the data even though the various protocols place important information such as machine addresses in different parts of the data stream. This conversion is often performed by gateways. Many routers offered today can provide the gateway function, which makes it more confusing when trying to identify the various pieces of the network. Understand what the gateway function is rather than what devices perform the function.

A gateway can also be the device used as an entry point into another network. The function of a gateway in this case is to provide security to the network, preventing unauthorized access. In the data world, this is referred to as a *firewall*. In the telephony world, these devices are called *gateways*. This can lead to some confusion when you are dealing with both types of networks.

4.2.8   Firewall

A firewall is a computer dedicated to providing security to the network. Any external networks being connected must pass through a firewall first, which screens out data from other networks based on criteria set by the network administrator. This is a software function that typically focuses on the address of the originator and the type of message being sent into the network.

Firewalls are extremely important in today's networks, due to the increase of hacker activities. They safeguard the resources within a network by denying access from outsiders, but they are not always foolproof.

4.3   CLIENT SERVER

First there was the mainframe, which used a centralized processor and centralized mass storage. All applications ran from the mainframe, accessed by terminals and later by PCs. As the PC market flourished, the LAN proliferated. However, LANs do not address the need for centralized databases, applications that are shared, or centralized mass storage (for access by everyone on the network). That is the purpose of client/server networks.

A client is an application or program running on the desktop. It provides the protocols and interfaces necessary to interact with an application stored on a server. A server is nothing more than a high-powered desktop computer (although today we see dedicated computers designed for use as servers). It is located on the network and can be accessed by all nodes on the network. Servers can be used to address a specific network application, or they can be multipurpose, dedicated to several network functions. Servers can be used as

- Communications servers, providing access to modems and/or other network connections

- Print servers, controlling access to a number of printers located on the network

- Internet servers, controlling Internet applications such as File Transfer Protocol (FTP) and the World Wide Web (WWW)

- Application servers, providing access to common applications used enterprisewide

- Database servers, storing and managing enterprise databases

It is somewhat ironic that the networking industry has reverted to centralization of some functions, when the very reason for using LANs is to use distributed processing.

However, some network functions must be centralized to be effective.

4.4 NETWORK OPERATING SYSTEMS (NOSs)

Network operating systems (NOSs) provide diagnostics and management in corporate networks. There are a number of vendors who provide these, including Sun, Microsoft, and Novell. The NOS usually provides protocols that manage e-mail transmissions, disk allocation, mass storage access, and monitoring.

There are many NOSs on the market today. Each is a proprietary solution with interfaces to more standardized solutions such as TCP/IP. In large enterprise networks (client/server), NOSs are needed so that network administrators can effectively monitor and manage the various resources on the network.

4.5 INTERNETS

Internets should not be confused with the public Internet. An internet can also be considered a private network used within a company to connect all of the various computers within the enterprise. An internet is two or more networks linked together, forming one ubiquitous network. The internet should be transparent to the end user and should interoperate (or *interwork*) with all other subnetworks. This presents a challenge when mixing equipment from many different vendors. This is why standards are so important.

The term *Internet* (capitalized) refers to the worldwide Internet that is used by many corporations and individuals today to link their individual networks.

Subnetworks are the individual networks within a larger network deployed by different corporations, service providers, and other large network users. A subnet may have a small number of nodes, or it may consist of other subnets (as is the case with many service providers). As a result, hierarchically, an internet may be several layers deep.

It is important to understand the relationships various subnetworks have with one another to understand the inner workings of TCP/IP. The devices used to interconnect the various subnetworks are routers and gateways.

A router receives data packets and forwards them through a port to another network or another part of its own network. A gateway works the same way as a router (from a TCP/IP perspective) but provides access to another network. A gateway can be considered an entry/exit point from one network to another. Gateways can also provide protocol conversion from one network to the next.

4.5.1 Autonomous systems

Internets are grouped into autonomous systems. An autonomous system is a group of networks joined together and maintained by a single authority. Autonomous systems are then linked to other autonomous systems by gateways. This provides a hierarchical approach to internets and simplifies the task of routing within an internet.

5

Ethernet

Ethernet was originally developed by Xerox and DEC as a way to interconnect their machines without the use of a mainframe network. The original Ethernet protocol was adapted by the Institute of Electrical and Electronics Engineers (IEEE), which made many improvements to the original design. The outcome was the Ethernet standard commonly used today, 802.3, which is called Carrier Sense Multiple Access with Collision Detection (CSMA/CD).

The IEEE also improved on the two protocols used in Ethernet and created another standard called 802.2. This standard deals with the actual packetizing of data and identifies the protocol structure, whereas 802.3 defines the standard used to prevent multiple computers from sending data at the same time (which results in collisions).

Ethernet uses a bus topology. The medium used in early Ethernet networks was coaxial cable. Thin coaxial is referred to as 10BASE2. In both cases, the 10 represents 10 Mbps, which is the bandwidth of Ethernet. The BASE indicates that the network is baseband, which means only one station can transmit at a time. The last number indicates the maximum distance supported by the medium. A 5 indicates 500 meters (m), and the 2 represents 200 m.

This has changed with the introduction of newer cables capable of transmitting data at higher speeds over thin copper wires twisted together (like those used in telephone networks). The difference between this new cable and conventional telephone cable is in the shielding used and the method used for twisting the cable pairs together. The tighter the twists, the faster data can be sent without interference.

These newer cables are rated for their transmission speeds and referred to as category 1, 2, 3, 4, or 5 cables. Category 5 cabling is the highest-rated cable available, supporting high data rates, and is the most commonly used cabling today. These cables also have separate cable pairs for voice transmission, allowing installers to use one cable to support computer and voice communications at each location.

In any bus topology, only one computer can send data at a time. If more than one computer is sending electrical signals over a copper wire, the electrical signals are going to "collide," causing the data to become corrupted and creating data errors. The IEEE defined a means for preventing this in the standard 802.3. What it defines is a way for computers to listen to the network, looking for an electrical carrier signal before transmitting data over the bus. The MAU performs this. If no carrier signals are present, the MAU begins transmitting its traffic over the network. This is not a foolproof technique, but it works most of the time.

If simultaneous transmission does occur between two nodes, collision detection (CD) is capable of determining that there was an error; it waits to see if there is any carrier (electrical current on the wire) and then allows the node to retransmit its data. This is an improvement over the original Ethernet standard, which did not have any collision detection or ability to determine if other nodes were transmitting.

In earlier Ethernet, a node sent a packet of data and waited for an acknowledgment. If an acknowledgment was not received within a few seconds, the node retransmitted the data. With CSMA/CD, the node will retransmit if collision is detected or if no acknowledgment is received.

Ethernet operates at the data link layer (as do all LAN protocols) (see Fig. 5.1). The functions defined by the Open Systems Interconnection (OSI) reference model for the data link layer do

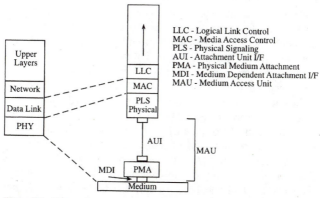

LLC - Logical Link Control
MAC - Media Access Control
PLS - Physical Signaling
AUI - Attachment Unit I/F
PMA - Physical Medium Attachment
MDI - Medium Dependent Attachment I/F
MAU - Medium Access Unit

Figure 5.1 Ethernet layers.

not sufficiently handle the needs of the LAN, so LAN protocols divide the data link layer into two layers, or sublayers. The sublayers used in Ethernet and most other LAN protocols are logical link control (LLC) and media access control (MAC).

5.1   MEDIA ACCESS CONTROL

The layer closest to the physical layer is the MAC sublayer. It deals with media access. It assembles the data into Ethernet frames when data is passed from the layer above (LLC) and prepares the data for transmission over the physical layer. The address of the node must be appended to the frame, and error detection/correction must be performed before transmitting the frame over the LAN.

When a node receives data, the MAC sublayer reads the address from the protocol header and determines if the address is its own. If it is, the MAC sublayer copies the packet, strips off the protocol header, performs error detection/correction, and passes the data to the next sublayer (LLC). The packet continues on the bus until it reaches the termination where the data is absorbed (like a sponge at the end of the bus).

Another function of the MAC layer is to add the frame check sequence (FCS) to the header. This is used to determine if the data sent was received without error. As with many error detection/correction methods, an algorithm is used to check the data received from the LLC. The results of the calculation are then entered into the FCS field of the header. When the data is received, the MAC layer at the receiving node performs the same calculation and compares the results with the value in the FCS field of the header received. If the values do not match, it is assumed that there was an error in transmission, and the receiving node requests a retransmission. The method used to request a retransmission varies from protocol to protocol. A particular Ethernet frame may be discarded for other reasons as well. For example, if the frame length is not valid, the frame is discarded. The MAC layer also performs this task.

There are two addresses in the MAC header. The source and the destination address identify the machine sending the Ethernet frame and the machine to receive the Ethernet frame (respectively). They are both 6 bytes in length and are hard-coded into the NIC cards of the sending and receiving computers.

The first bit of the address signifies an individual machines address (if it is 0) or a group address (if it is 1). A group address can be assigned to a number of machines on the same network. For example, a work group such as engineering may have a group address for hardware engineering and a group address for software engineering. The system administrator, who must also configure group machines as part of a group, determines this. An address of all 1s is a broadcast address destined for all machines on the network. Figure 5.2 shows the Ethernet frame and its parts at the MAC level. The LLC layer will of course add additional information.

The preamble is a pattern of bits consisting of alternating 0s and 1s. The pattern is determined by the sending node and must not replicate any patterns sent within the data portion of the frame. The start frame delimiter then follows this, which is a predefined pattern that indicates the beginning of the Ethernet frame. This is used together with the preamble. The start frame delimiter is 1 byte in length.

Following the preamble and the start frame delimiter are the destination and source addresses. All machines on the network must have an Ethernet address. It is probably worth noting here

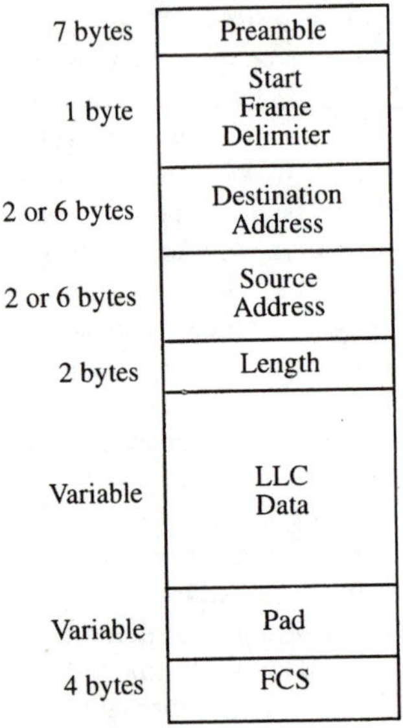

7 bytes	Preamble
1 byte	Start Frame Delimiter
2 or 6 bytes	Destination Address
2 or 6 bytes	Source Address
2 bytes	Length
Variable	LLC Data
Variable	Pad
4 bytes	FCS

Figure 5.2   Ethernet frame.

that the Ethernet address is used only within Ethernet and is of no significance when a higher-layer protocol such as TCP/IP is used. TCP/IP uses its own addressing, but it does not identify the machine address of nodes within the destination network. Addresses at higher layers identify users and applications rather than physical entities.

The length field of the frame identifies the length of the data field. This is necessary because the data field is a variable-length field. The data that is inserted in this part of the frame includes any data appended by protocols. For example, the LLC header can be found in the data portion of the MAC layer. You will find in the LLC header a data portion as well, which will include data appended from higher-layer protocols as well as the user data. This is an important concept to understand. As we look further and further into the protocol stack, you will see that each protocol used appends its own information into the data portions.

The pad field is used to maintain a consistent length to the overall Ethernet frame. The pad is nothing more than all 0s inserted after the variable-length data field to ensure consistency in Ethernet frame lengths despite variable lengths of data.

5.2 Logical Link Layer

There are many different types of LLC headers; their use depends on the type of service being provided by Ethernet. Three types of services are offered by Ethernet 802.3: unacknowledged connectionless service (type 1), connection mode service (type 2), and acknowledged connectionless service (type 3). The sending node determines the type of service to be provided.

This is the fundamental difference between the IEEE 802.3 standard and the earlier Ethernet 2.0 standard. The Ethernet 2.0 standard only supports unacknowledged connectionless service and combines the MAC and LLC functions into one layer. Unacknowledged connectionless service is the least reliable method of sending data. Connectionless means that prior to sending data, the protocol makes no attempt at establishing a logical session with the destination node. The data is placed in frames and transmitted to the destination node unannounced.

With this level of service, there is one basic message type (or data unit); unnumbered information (UI). Unnumbered means that there are no sequence numbers sent with each associated data unit. The frame must contain all of the information necessary to process the data at the destination node. Again, there is no guarantee that the destination node will ever receive the data because there are no data acknowledgments sent by the receiving node. LLC management uses this type of service to exchange identification data with each node. Besides the machine address, the service access point (SAP), which identifies the various operations in the system, is also provided. This frame type is designated exchange identification (XID).

There is also a test message frame, used by protocol analyzers to send "loopback" frames over the Ethernet when testing. This frame type is also classified as unacknowledged connectionless service.

Connection mode service is a connection-oriented protocol. Connection-oriented protocols negotiate a session with the destination node prior to sending any data. A session is purely logical and does not represent a physical connection between nodes. Several sessions can run over one physical connection at a time, all of them identified by the various addresses we have already identified.

The logical session guarantees delivery of all data by acknowledging every data frame received. Sequence numbers are commonly used in connection-oriented protocols, allowing the destination node to verify that it has received the data units in the same sequence that they were transmitted, preventing out-of-sequence data errors.

The first phase in connection mode service is to establish a connection. Once the logical connection is established, data is transferred between the two nodes. When all of the data has been transmitted, a disconnect is requested (by the originating node), and the resources allocated to the connection are released for another session.

Several different frame formats are used with connection mode service. The information frame is used for transmitting user data. Figure 5.3 shows the frame format for an information data unit.

The N(s) and N(r) fields are where the sequence numbers are found. The N(s) field contains the sequence number of the frame being sent, and the N(r) field contains the sequence number of the last received frame. This allows any node to verify that data is received in the correct order and provides a means for acknowledging all received frames in one transmission.

The supervisory frame is used for sending acknowledgments when there is no data to send (see Fig. 5.4). If a node has

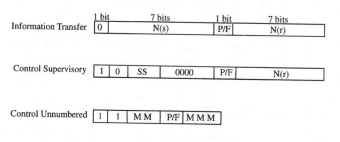

N(s) - "Now sending"
N(r) - "Need to receive"
S - Supervisory function bit
M - Modifier function bit
P/F - Poll/Final bit

Figure 5.3   Ethernet information frame format.

| Control Supervisory | 1 | 0 | SS | 0000 | P/F | N(r) |

N(r) - "Need to receive"
S - Supervisory function bit
P/F - Poll/Final bit

Figure 5.4   Ethernet supervisory frame format.

received data but has no data to send in return, it can use the supervisory frame to send the acknowledgment (there is no user data field in the supervisory frame). This frame format is also used for flow control.

Flow control is used to stop data flow from the originating node. This consists of receiver ready and receiver not ready message types. The reject frame is used to request a retransmission if the frame is considered invalid. Invalid is different from a frame received with errors. The frame may have been of the wrong length, or it may be so corrupted that the header cannot be deciphered.

The unnumbered frame format is used when initially establishing a connection between two nodes. It is also used to send a disconnect message to a node, releasing a logical connection. The format used in unnumbered frames is shown in Fig. 5.5.

5.3   ACKNOWLEDGED CONNECTIONLESS SERVICE

Acknowledged connectionless service uses acknowledgments to guarantee delivery of each data unit. Every data unit is acknowledged as it is received, which is different from connection mode. In connection mode, the receiving node can acknowledge a whole contiguous series of data units. Not acknowledging every data unit as it is received saves on

Control Unnumbered | 1 | 1 | M M | P/F | M M M |

M - Modifier function bit
P/F - Poll/Final bit

Figure 5.5 Ethernet unnumbered frame format.

resources and cuts down on the amount of traffic sent over the network. This is known as asynchronous acknowledgment.

In acknowledged connectionless service, every frame must be acknowledged as it is received. If no acknowledgment is received for a frame, the originating node assumes it was lost during transmission and retransmits the data unit again. This of course adds traffic to the network. This type of service is not used often, but it can be useful for some types of data (such as e-mail).

5.4 FAST ETHERNET

There is a new standard for "fast" Ethernet. Fast Ethernet supports transmission rates up to 100 Mbps over coaxial cable or category 5 twisted pair. The functionality of the protocol is somewhat the same, although some of the protocol has changed to support the higher data rates. Fast Ethernet is very competitive with FDDI and Token Ring networks because it can use the existing network infrastructure.

6

Token Ring

IBM first proposed Token Ring in 1969. At the time, Ethernet was limited to 10 Mbps. Token Ring was capable of supporting 16 Mbps, which helped its popularity. Today, this is no longer an attraction since Ethernet is capable of supporting 100 Mbps (Fast Ethernet). The IEEE later adopted Token Ring as standard 802.5. There is also a Token Bus standard (802.4), which is similar to Token Ring except it uses a bus topology instead of a ring topology.

The principal difference between Token Ring and Ethernet is in the topology and the use of a "token." Whereas Ethernet uses a bus topology, Token Ring relies on a ring topology. Many of the same limitations of Ethernet apply to Token Ring, including the fact that only one node on the network can transmit at a time. Remember that with bus topology, the data is broadcast over the bus (in both directions) by an originating computer. Terminators connected to the ends of the bus absorb the data. In Token Ring networks, the data is handled differently. The originating computer sends the data over the ring to an adjacent computer. The data is read by the adjacent computer, and if the computer determines that the destination address is not its own address, it regenerates the data packet without modification. However, if the destination address is its own, the packet is modified to indicate that it was received and is copied into memory. Then it is sent out onto the ring again, where it travels around the ring until it reaches the originating computer. The originating computer then reads the fields of the packet to verify it reached its destination without error. The originating computer is responsible for removing the data packet from the network and generating a "token" packet to indicate the network is now available for transmission.

To resolve data collision, Token Ring uses a more reliable method than Ethernet does. A small data frame, referred to as the *token,* is transmitted around the network. A node cannot transmit data until the token reaches the node. It can then hold the token and transmit its data. Once it has completed data transmission, it releases the token for the next node.

Another principal difference that is important to understand is that every node reads data frames as they are received from the network and then retransmits them. This is different from Ethernet, where the data frames are transmitted over the bus, and each node reads the address of the frames. The data frame is really broadcast over the bus. In Token Ring, the data frame is passed from one node to the next, each time being re-created by the node.

Because the data frame is re-created at each node (or more accurately regenerated), repeaters are usually not needed in

Token Ring networks. Every node is a repeater, allowing for more distance in the network. However, this does not mean that Token Ring can handle large networks better. All networks become inefficient when too many nodes are attached to the same network segment.

Because the data frame must pass through each node in the network, delay becomes an inherent problem of Token Ring networks. The more nodes in the LAN segment, the more nodes the data must pass through. Each node can add as much as a 1-second (s) delay to data transmission. This is one disadvantage of Token Ring networks.

Like Ethernet, Token Ring divides layer 2 into two sublayers: the MAC and the LLC sublayers. Because LAN protocols tend to be topology-specific, there are obviously differences between the Ethernet MAC/LLC and the Token Ring MAC/LLC. The physical layer is also very different from Ethernet. As mentioned earlier, each node is connected to a ring. This means that each node is connected to its neighboring node (rather than to a bus). There is only one connection, made using a connector. Earlier versions of Token Ring required that the nodes remain powered on at all times; otherwise they would cause the ring to fail (because they were responsible for regenerating the data frame to the rest of the ring). This was later resolved through the NIC, where a shunt was placed to allow nodes to be powered down without interrupting the network. The shunt acts as a bypass when the computer is powered down.

If we look at the frame format of Token Ring (Fig. 6.1), we can see some similarities to Ethernet and other protocols. There is a delimiter to identify the start of the frame (much like the start frame delimiter in Ethernet) and an end-of-frame delimiter.

SD	AC	FC	DA	SA	INFO	FCS	ED	FS

SD - Start Delimiter
AC - Access Control
FC - Frame Control
DA - Destination Address
SA - Source Address
FCS - Frame Check Sequence
ED - Ending Delimiter
FS - Frame Status

Figure 6.1   Token Ring frame format.

The fields in the front of the frame are referred to as the start-of-frame sequence (SFS). The ending fields are referred to as the end-of-frame sequence (EFS).

The token that is passed around the ring consists of only the start-of-frame and end-of-frame sequence fields. There is nothing in between. As seen in Fig. 6.1, the only fields in the SFS and the EFS are the starting delimiter (SD), access control (AC), and the ending delimiter (ED).

The start-of-frame delimiter starts with 2 bits followed by a 0, followed by 2 bits and three 0s. This pattern cannot be repeated in the data field anywhere because it may then be mistaken for a new frame. This means that the pattern is variable, depending on the data contained in the frame.

The access control field is used to identify and control access to the ring. The first 3 bits are called primary bits, and they signify the priority level of the token. This is used by each node to determine whether or not it can hold the token or if it has to continue passing it along. Each node has a priority as well and cannot hold the token unless its priority is equal to or greater than the token priority.

This sounds complicated (and it can be), but the basic idea is to use a priority scheme to prevent nodes from taking the token out of turn. For example, let us say that node A is transmitting, and node C wishes to transmit next. While waiting for node A to finish transmitting and generate a token, node B suddenly decides it wants to begin transmitting. Since node A is connected to node B, node B would be next in line to receive the token, even though node C has been waiting longer. If node C could identify itself as having a higher priority than node B, the token would be passed through node B to node C.

The next bit in the access control field is the token bit. This is used to identify a frame as a token. If the value in this field is a binary 0, the frame is a token frame. If the value is a binary 1, the frame is a data frame.

Following the token bit in the access frame is the monitor bit. This bit is used for managing the network. There must always be one node on the ring acting in monitor mode. This is the node that will be responsible for generating the first token when the network is first activated. The monitor node will also be responsible for removing from the ring data that never reached its destination

The next 3 bits are called reservation bits and are used along with the priority bits. This allows nodes on the network to reserve the next token for them. As data frames are passed around the ring, these fields in the access control field are altered as they pass through each node on the ring.

The next field is the frame control field, which identifies the type of frame being sent (MAC or LLC). This is followed by an ending delimiter field, which carries a unique pattern of bits that is not duplicated in the data portion of the data field. This delimiter identifies the end of a frame.

The last field in the frame is the frame status. The node identified in the destination address field uses the frame status field. The receiving node of a frame identifies whether or not it recognized its address in the destination address field and also signifies whether or not it copied the frame into memory. If there was an error of some sort, the destination node could signify that it recognized its address but did not copy the frame into memory. This is the way that Token Ring asks for retransmission of a frame. The originating node then reads this field when the frame comes back around to it again, and if the frame-copied bits are negative (did not copy the frame), it will retransmit the frame.

Now that we have identified the fields used in Token Ring, let us look at how data is passed around the ring. As a token passes around the ring, a node wishing to transmit data waits until the token passes by it and then seizes the token from the ring. The token bit is changed to indicate a data frame, and the MAC layer appends the remaining fields.

The node then transmits the data frame over the ring. If there is a lot of data to be sent, the originating node can continue transmitting data frames until it is either finished or a token-holding timer expires. The token-holding timer prevents any one node from "hogging" the network time by continuously transmitting and never releasing the token. As the data frames are passed around the ring, each node on the ring reads the data frame, looking at the address fields and performing error detection to determine if an error has occurred between it and its adjacent node. If an error is detected, the error bit found in the ending delimiter field is set.

The data frame continues its journey around the ring until the destination node recognizes the destination address. The destination node must then determine if there is enough buffer space to copy the received frame. If there is enough room in its buffers, it changes the status bits to indicate that the address was recognized and that the frame was copied. The frame is then retransmitted over the ring.

The originating node is responsible for removing its data frame from the ring. When the data frame reaches the originating node, it is removed from the ring and the token is then re-created and transmitted. This sequence of events is repeated for every data frame sent on the network.

Each node on the network can also set a priority bit in a passing data frame, which is used in combination with the reservation bit to ensure that the node receives the next token transmitted. This mechanism is necessary because it is possible that an upstream neighbor that has been waiting a shorter period of time to transmit data could seize a token.

As data frames are passed around the ring, nodes wishing to transmit set the reservation bit to a priority higher than what is currently set in the priority field. When the originating node has completed transmission, it sets the priority bit one level higher and transmits a token. Stations that have indicated a lower priority cannot seize the token. For example, let us say that node A is transmitting data. Node D reads the data frame and the priority bit, which is set to 5. Node D then sets the reservation bit to 5. When node A has finished transmission, it generates a token and sets the priority of the token at 5.

Other nodes on the ring cannot grab the token because they will have a lower priority than 5. When node D receives the token, it begins sending its data frames. Once node D has completed transmission, it repeats the cycle we just described, transmitting a token with a priority of 5 (or higher if another node has set the reservation bit higher). When the token passes by node A, node A decreases the priority back down to a lower priority so that it can be seized by upstream neighbors.

This is a very brief overview of the reservation system, which is actually a bit more complicated than described. The main point is that each node has a fair shot at receiving the token and sending data. This requires the use of the priority and the reservation bits. You may have figured out one flaw in this protocol. The originating node is responsible for removing its data frame from the network. But what happens if the originating node shut down shortly after transmission? The data frame can pass around the ring forever because the originating node is not present anymore. Even if the originating node were powered back up, it would not recognize the data frame as its own because all of its buffers would have been reset.

To prevent this from happening, one station always acts as the monitor station. The job of the monitor station is to read each data frame and determine if the frame has passed by it once already. The monitor station knows this because the first time a frame passes by the monitor station, it sets a monitor bit to 1. Any frame passing the monitor station with a monitor bit set to 1 has already passed by the monitor station once. The monitor station then removes the data frame from the network.

Any node can be the monitor station. When the ring is first initialized, the first station to be powered up sends an Active-Monitor-Present frame, indicating its wish to be the monitor station. If a node has already identified itself as the monitor station, when it receives the Active-Monitor-Present frame, it automatically goes into standby mode, relinquishing its status as network monitor.

7

Fiber Distribution Data Interface (FDDI)

FDDI was developed by the IEEE and adopted as an ANSI standard through the X3T9.5 committee. Although this is a relatively new standard (first introduced in 1992), work on it has been under way for some time. The concept for a fiber optics network utilizing a ring topology was first introduced in 1982.

FDDI is similar to Token Ring in that it uses a token to identify an idle network, eliminating the possibility of data collisions. Where FDDI differs is in its handling of the token. FDDI uses a timed token, allowing nodes to hold tokens for a predetermined time before they are forced to relinquish control of the token. FDDI also uses a dual ring topology. This ring is counterrotating, which means that data is passed in both directions. This adds to the integrity of the network. As seen in Fig. 7.1, each node has two inputs and two outputs, each moving in opposite directions. If the ring breaks between two nodes, the nodes pass data back to the other ring, forming one ring rather than two.

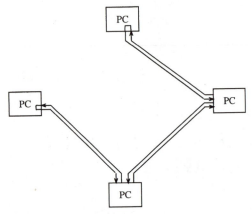

Figure 7.1   Counterrotating FDDI Ring.

FDDI relies on other protocols above its own MAC layer, such as Ethernet MAC (802.2). In reality, where FDDI works best is as a backbone network used to move data from one LAN to another. FDDI provides 100-Mbps bandwidth, easily handling the capacity of Ethernet and Token Ring networks that may be attached. This is important for any backbone technology because the aggregate bandwidth from multiple networks may overwhelm the backbone.

Instead of going into the specifics of the FDDI protocol (such as frame types and protocol specifics), we will concentrate on how it operates. As mentioned earlier, FDDI uses a timed token mechanism. What this really means is that a node is allowed to hold a token for a limited amount of time and then it must release it. This is somewhat different from Token Ring because a node is allowed to hold the token for a longer period of time in Token Ring.

What is of significance in FDDI is that the token is released immediately after the data frame is transmitted. The originating node does not wait for the data frame to return from its trip around the ring. This means that multiple nodes can be transmitting over the ring, as long as they wait until the token is received. There is no danger of data collision because nodes must wait until a token has passed before transmitting. There is the possibility that multiple data frames will exist on the ring at the same time. As with Token Ring, the originating node is responsible for removing its own data frames from the network when it has passed one complete revolution.

FDDI also uses a distributed ring management function. Each node is responsible for communicating with its neighboring upstream nodes. Status information is constantly shared with adjacent nodes, allowing any node in the network to initiate ring initialization, isolate a fault, and recover from ring faults.

There are many other advantages to FDDI. Tables 7.1 and 7.2 compare FDDI with Ethernet and FDDI with Token Ring and show that FDDI is capable of outperforming either protocol. Keep in mind, however, that there is now a 100-Mbps version of Ethernet that can be a stiff competitor with FDDI.

To understand more about FDDI, we will look at the various sublayers defined in the FDDI standard and will review their responsibilities. We will not go into great detail about how the various sublayers work but will look at what they do.

The FDDI standard defines two layers which compare with layer 1 in the OSI model. The physical media dependent (PMD) layer is the lowest layer, and it specifies all hardware

TABLE 7.1 Comparison between Ethernet and FDDI.

	FDDI	Ethernet
Data rate	100 Mbps	10 Mbps
Maximum frame	4500 bytes	1518 bytes
Encoding	80% efficient	50% efficient
Distance between nodes	2 km	0.5 km
Maximum length	100 km	2.8 km
Maximum nodes	500	1024
Topology	Dual ring	Bus
Access	Token	CSMA/CD

and software interface operations. This includes the transmission of data over optical fiber, the connectors, services to the upper layers, and signal waveform and code requirements.

PMD supports single- and multimode fiber optic cable. The difference between the two lies in the way light passes through the fiber. In single-mode fiber, the inner core of the fiber is narrow, and the inner surface of the fiber is less reflective than in multimode. The actual core is hollow, allowing light to pass straight through. A laser is used as a light source, shooting a light beam in a straight pattern through the inner core. Only a single beam can be used because other light beams would interfere with each other.

In multimode fiber, the light beams are aimed at angles. The inner core of the fiber is somewhat wider and more reflective.

TABLE 7.2   Comparison between Token Ring and FDDI.

	FDDI	Token Ring
Data rate	100 Mbps	4 or 16 Mbps
Maximum frame	4500 bytes	No limit
Encoding	80% efficient	50% efficient
Clock	Each node	Master node
Maximum length	100 km	10 km
Maximum nodes	500	50
Topology	Dual ring	Single ring
Access	Token	Token

The light actually bounces off the inner walls of the fiber core at specific angles. Other transmissions can be represented by additional beams at different angles, allowing multiple beams to exist at the same time in the same fiber.

Several different connector types are used in FDDI. An MIC connector is used to connect two rings to one node. Each node has two MIC connectors and the transmissions are split between the two connectors for optimum redundancy. One pin on the MIC connector is for data coming in on one ring, and the other pin is used for data going out over the other ring. With two connectors, a node can be ensured that it will always have a connection to one ring or the other even if there is a fiber cut.

An ST connector is used when a connection to only one ring is desirable. Each node using an ST connector has two connectors, one for data coming in on the ring and the other for data going out on the ring. These are used for connecting nodes to concentrators.

Nodes can be connected in one of two ways. A node can be a dual attached station (DAS) or a single attached station (SAS). In a DAS configuration, a node is connected to both the primary and the secondary ring through an MIC connector. This is not always possible, depending on the area being cabled. In some cases, it may be more desirable to run a single set of fibers to a node and use a concentrator (to feed a number of nodes). When a concentrator is used, a special cable is used to connect both rings to it. Each station is then connected using single fibers. Only the primary ring is connected to each node. This configuration is referred to as SAS.

The physical (PHY) layer protocol lies directly above PMD. PHY provides the encoding and decoding of data before it is transmitted over the fiber by the PMD sublayer. It is also responsible for transmitting and receiving data, clock rate and synchronization, the various line states, and services to the upper layers.

Line states indicate what is happening over the fiber. There are several possible states:

- Quiet line state (QLS)

- Master line state (MLS)

- Halt line state (HLS)

- Idle line state (ILS)

- Active line state (ALS)

- Noise line state (NLS)

QLS indicates a loss of signal from PMD, which usually indicates a break in the fiber or a missing connection. MLS is used as part of the connection establishment phase. Before data can be transmitted over the fiber, it must be established that all connectors are functioning and the ring is operational. The node will go into several line states during this process to ensure ring integrity before transmitting. HLS is also part of this process. ILS is used to establish and maintain clock signals over the fiber. It is used as part of the connection establishment phase as well as between frames. ALS is used to indicate a connection has been established. NLS is used to indicate that there is noise on the fiber and there may be a faulty connection.

The MAC layer identifies the frame types being sent and also defines frame formats and sizes. When the network is idle, network management messages are passed around the ring to maintain integrity of the network. This includes status messages sent from one node to the next. The MAC layer also performs ring timing, fault isolation, and frame removal.

Several timers are used in FDDI. They are used to determine how long it will take for a message to traverse one revolution around the ring. This information is also used to maximize the available bandwidth on the ring. The results from these various timers are used to calculate how long a node is allowed to transmit before it must surrender a token, allowing other nodes to transmit. This mechanism is a bit more sophisticated than the timer used in Token Ring.

The token rotation timer is used to time how long it takes a token to make one revolution around the ring. Station management (SMT) sends this when the ring is first initialized (during what is referred to as the claim process). Each node times how long it takes for a token to pass around the ring. A claim frame is then sent, with the value from the token rotation timer.

As the claim frame passes around the ring, each node examines the value of the timer and compares it with its own. If the node has a lower value than what is provided in the claim frame, it changes the value in the frame and regenerates the claim frame. This process is repeated at every node. When the claim frame has passed around the ring once, each station again looks at the value in the frame to determine if it matches its own timer value. If it does, the node "wins" the claim process and generates the first token.

The token-holding timer is used to determine when a node must surrender the ring and generate a token. This means the node must stop transmission to allow another node to transmit data. If a token is received sooner than expected (according to the token rotation timer), the difference is then allocated to the token-holding timer, allowing the node to continue transmitting.

The valid transmission timer is reset every time a valid frame is received. If the timer expires, the node then initializes ring recovery procedures. This is used to manage the integrity of the ring and provides a mechanism for ring error detection.

FDDI offers an alternative to broadband services for large networks such as campuses and hospitals. It makes an excellent backbone for interconnecting multiple networks to form one contiguous network. Based on fiber optics, it also offers a more reliable and error-free transmission medium than copper.

8

Transmission Control Protocol/
Internet Protocol (TCP/IP)

TCP/IP started as a robust network called ARPANET that was used by the government. The military wanted a network that would span the nation and be capable of maintaining connections even if a city had been devastated during war. It became popular among scientists and researchers and soon found its way onto U.S. campuses involved in research projects for the government. Naturally, students who had access to the ARPANET looked for ways to continue using it. When the ARPANET was later made public and became the Internet, many companies began looking for ways to get connected.

The reason for all of the interest is because TCP/IP provides a number of protocols and solutions that meet just about every corporate networking need. To top it off, TCP/IP is a very robust yet efficient network solution, making it cost effective for businesses who need to link various locations with corporate headquarters.

TCP/IP is not just for linking locations with one another. It's about linking millions of users with one another. This network solution allows businesses to communicate with one another, share e-mail across the globe, send files from one city to another, and exchange ideas through special forums called *newsgroups.*

The concept of the Internet has moved indoors. Many are now finding that the same suite of protocols used to connect the world work just as well within the walls of the corporate office. For this reason, companies are now looking to TCP/IP to support their own intranets: TCP/IP networks that are used internally to share files, e-mail, and data exchange.

TCP/IP is a suite of protocols. These protocols fall within various layers of the protocol stack. To understand the TCP/IP protocol stack, one should understand the interactions between layers.

Subnetworks are managed at the physical and data link layers. Although these layers are not part of TCP/IP, they do interact with the protocol stack. For example, to reach an IP address, the IP address must first be translated (or resolved) into a LAN machine address. If the IP address is part of a subnetwork, the subnet mask must be determined and the address resolution based on the results of the subnet mask.

Internetworking is managed by IP at the network layer. IP does not support error control, so it relies on another protocol for this function. The Internet Control Message Protocol (ICMP) provides error correction and flow control for IP. ICMP, although a user of IP, is still considered part of the network layer (ICMP information is encapsulated in IP packets, making it a user of IP).

TCP and the User Datagram Protocol (UDP) are considered service provider protocols and reside at the transport layer.

TCP is a connection-oriented protocol, whereas UDP is a connectionless protocol. Both rely on the services of IP but do not require IP. For example, TCP and UDP can be transported over X.25 or Frame Relay services (both of which reside at the network layer).

The applications service layer consists of a number of protocols such as FTP, TELNET, and Network News Transport Protocol (NNTP). These are not really applications themselves but are protocols that interface with the various applications that are necessary to use these services. They provide the communications to remote devices but do not provide the user interface to interact with the various remote services. All of these protocols encapsulate data into envelopes referred to as *protocol data units* (PDUs). There are many different labels used for these PDUs at various layers. In this chapter, *segment* will be used to describe PDUs from the transport layer (such as TCP) down to the network layer. In other words, when a protocol passes data from TCP to IP, the data unit is referred to as a segment.

Datagram is used to refer to PDUs passed from the network layer down to the data link layer (as in from IP to Ethernet). Datagrams sometimes also refer to packets in connectionless protocols, such as UDP. Once a data unit has passed through the various layers and is sent to the physical layer, it is considered a *frame*. Once the data unit has been passed over the network, it is referred to as a *packet*.

To interact with other parts of a host (software modules), protocols must interact with interfaces provided by the operating system. The operating system provides ports as entries to applications. As a data unit is passed to the application layer, the operating system provides a connection to the application by way of a logical port. The connection is established and maintained throughout the transaction period until data segments have been terminated.

A socket identifies an endpoint communications process. For communications to pass from one machine to another, a port must be connected and a socket defined. Internet ports are usually predefined (0 to 255) for well-known applications (such as FTP and NNTP). Undefined ports are provided as well, allowing operating systems to define their own ports when necessary.

Now that we understand some of the terminology used with this technology, let us look at some of the advantages of TCP/IP. When data is sent from an application down to the transport layer, the data may be too big to fit into one data unit. TCP provides a service called *fragmentation and reassembly* to handle this problem. The data is divided into evenly sized data units and is then passed to the network layer for further processing.

At the network layer, the individual data units may require further fragmenting. There is nothing wrong with this practice since protocols work within a peer-to-peer relationship. Data fragments created at the transport layer cannot be processed at the network layer; they must be passed to the transport layer before the fragments can be reassembled and processed. So fragmenting at various layers does not pose a problem.

When the data unit is passed over the network, it must pass through routers and gateways to reach its destination. It is possible that a data unit may pass through a network that will not accept the size of the IP fragments, and the data units will have to be fragmented further. This is not a problem for IP or TCP, which manages fragmented data at both the network and the transport layers.

When a host receives fragments, the IP layer looks for the other parts of the data. Timers are used to determine when fragments are considered lost, and when a timeout occurs, the received data is thrown out and an error message is sent to the source to generate retransmission of all of the fragments.

In addition to this handling of data fragments, multiple addressing conventions can be supported between various subnets. Different routing methods can be used within the various subnets with absolutely no effect on the end-to-end transmission itself. For example, a message may be passed to a subnet using source routing (the source defines the path to take to reach a destination), even though the source originated the message using non-source routing. In short, a message does not have to follow one method of routing all the way through the network. Each portion of the network can use any routing mechanism it wants without affecting the delivery of the original data unit.

By the same token, various services can be provided from one subnet to the next. A data unit can be originated using the connection-oriented services of TCP, but along the way a subnet can use connectionless UDP to pass the message through its own network as long as TCP services are used at the destination.

There are many other advantages to using TCP/IP in an internetwork environment. In short, TCP/IP was designed to support data communications through a number of nonrelated networks, ensuring reliable delivery of data even when networks fail along the way.

8.1   INTERNET PROTOCOL (IP)

The IP resides within layer 3 (network layer) of the OSI model. It provides end-to-end transport of data units through internets using connectionless services. Being connectionless, IP does not provide reliable data transfer, but this is not an issue if the upper layers provide reliability and error control.

An IP host must encapsulate data into IP headers, which are then passed to the data link (such as Ethernet). The protocol at the data link layer then encapsulates the IP header with the data into its own data unit (the datagram). The datagram is then passed down to the physical layer, where it is passed over the network as a serial bit stream (with possible encapsulation again, depending on the technology used).

For data to leave the local network, it must be sent to a router. Routers are network layer devices and are capable of processing the Ethernet and the IP headers. If the data is to be passed to another network, the Ethernet (or data link) header is stripped from the data, and the IP header is then processed.

Before transmitting the data over a port to the next network, the router must create a new IP header and place the data (consisting of the TCP header, possibly an application header, and user data) into the IP header. The datagram is then given to the data link layer (which may now be X.25, ISDN, Frame Relay, or even Switched 56), and the whole process is repeated.

IP does have its limitations, the biggest being the number of addresses available. There is a severe limitation in the number of addresses that IP can support. This issue has brought about the need for a replacement to IP. Internet Protocol Version 6 (IPv6) provides a 16-byte address rather than a 4-byte address.

The primary function of IP is to provide routing information for data being transported through internets. Any error control is provided by the ICMP, which resides at layer 3 as well. This protocol does not provide error control but merely reports errors to the originating hosts.

IP is not a requirement for TCP, which can use almost any network layer protocol for delivery as long as the protocol is capable of providing routing services and supports the interfaces between the two layers. Remember that layering allows various layers of a protocol to be changed without affecting the layers above or below it.

8.1.1 IP header

Figure 8.1 depicts the IP header and its fields. The first field in the IP header is the *version* field. This is used to identify which version of IP was used to create the header. This is important in internets because not every network is running the same version of a protocol. If the IP header was created in a network using the latest version of IP, it may contain information not recognizable by an older version of IP. When this occurs, the receiving network (running the older version of IP) knows to ignore unrecognizable fields because the version field indicates a version newer than its own. This is valuable information for an internet and can be found in a number of the TCP/IP protocols.

The version field is followed by the *header length* field, which provides the length of the header itself. The data portion is not indicated here. One would think that the IP header would always be the same length, but there are some variable options that can be included in the header. The length is measured in 32-bit units (or words). A 1 indicates a header length of 32 bits, whereas a 2 indicates a header length of 64 bits.

Following the header length field is the *type of service* field (see Fig. 8.2). In other protocols, this is often referred to as QoS. It stipulates the level of service the data requires. This field contains the four values discussed below.

IP Header

Version	Length	Service Type	Total Length	
Identification			Flags	Fragment Offset
Time to Live		Protocol	Header Checksum	
Source IP Address				
Destination IP Address				
IP Options (optional)			Padding	
Data				

Figure 8.1   IP header format.

PPP	D	T	R	0	0

PPP = Precedence
D = Delay Attributes
T = Throughput Attributes
R = Reliability Attributes

Figure 8.2   Type of service field.

Precedence is used to assign a level of priority to a data unit. It can be used for congestion management (lower priority data units could be discarded whereas higher priority data units are allowed to pass) and flow control. Not all networks implement this parameter. The precedence field consists of a 3-bit code that indicates the type of precedence assigned. The values are shown in Table 8.1.

The *delay* field indicates whether a delay should be applied when sending the packet. The standard does not indicate how much of a delay should be used when this bit is set to 1. There are two possible values for this field, normal and low delay. A 1 indicates low delay.

TABLE 8.1 Type of service values.

Precedence	Reliability Attributes		Throughput Attributes		Delay Attributes	
111 Network control	0	Normal	0	Normal	0	Normal
101 Critic/ ECP	1	Best possible	1	Best possible	1	Best possible
011 Flash						
001 Priority						
110 Internetwork control						
100 Flash override						
010 Immediate						
000 Routine						

The *throughput* field can be used to indicate that high throughput should be used with the particular data unit. For example, if the data unit was generated by a real-time application (such as an interactive game), the application may request a speedy delivery of the data units, requiring high throughput. The possible values are normal or high.

The *reliability* field is used in a similar fashion, indicating whether or not this data unit requires high reliability or normal service. If high reliability is indicated, it may be necessary to apply additional services at the upper layers to provide a high-reliability transmission, or it may mean that the data unit should not be routed over certain routes that may not provide a particular level of reliability.

The standard does not define how these bits should be implemented within an internet. It is left to the network provider to determine how these should be implemented within its own network. The protocol does provide the mechanisms for various service types if network providers choose to use them.

Following the type of service field is the *total length* field, which provides the length of the header and the data field (which would consist of the TCP/UDP header and the user data). The maximum length data unit at the IP level is 65,535 octets. This parameter allows nodes to determine the length of the data field by subtracting this value from the header length.

The identifier field is a 16-bit field used to correlate data unit fragments. When a data unit is fragmented, a number is assigned by the source to the fragments so that the receiver can match the IDs and reassemble the packet. The ID for associated fragments is the same, so the receiver can determine which fragments belong to each other.

A 3-bit *flag* field is used to qualify data units for fragmentation and to identify the last fragment in a series of fragmented data units. The first bit in the 3-bit field is always set to zero. The second bit is used to identify whether or not fragmentation is allowed for a data unit. This bit can be used when routing a data unit through a network and you want to prevent intermediate networks from fragmenting the data unit. For example, a data unit may be sent in whole but may be too large for an intermediate network to pass. The intermediate network would then fragment the data unit. This bit would prevent the intermediate network from fragmenting the data unit and would possibly force the data unit to take a different route.

When data units get fragmented, the protocol must identify where each particular fragment belongs in the reassembled data unit. There are a number of ways protocols accomplish this. In IP, the fragment offset field is used to identify where a fragment should be placed in relationship to the whole data unit. This is then used at reassembly by the receiving node.

It is important to note here that IP does not "hand off" the single fragments to TCP and expect TCP to provide all reassembly of fragments. IP may have fragmented data units several times during its route to its final destination, making it impossible for TCP to reassemble the data unit. IP provides reassembly of fragmented data units at the IP level, and TCP provides reassembly of fragmented data units at the TCP level.

The *time-to-live* field is actually a hop counter. Each time the data unit traverses through a router, the router decrements this field by 1. The field originates with some value (maximum is 15, because 16 is considered as unreachable). The value is determined by the originating network, depending on the quality of service required by the data unit. When the time-to-live field reaches a value of zero, the data unit is discarded. This is used to prevent circular routes with an internet. For example, a cluster of routers may be passing a particular data unit around in a circle because of some network failure. When a router reads the time-to-live parameter and sees the value has reached zero, the router will immediately delete the data unit and pass an error message (using the ICMP protocol) to the originator.

The *protocol* field identifies the protocol contained in the data field. A number identified in the standards identifies each protocol related to TCP/IP. Table 8.2 contains the numbers assigned by the standards for well-known protocols.

TABLE 8.2   Protocol field values.

No.	Protocol
1	Internet Control Message Protocol (ICMP)
2	Internet Group Management Protocol (IGMP)
3	Gateway-to-Gateway Protocol (GGP)
6	Transmission Control Protocol (TCP)
8	Exterior Gateway Protocol (EGP)
9	Interior Gateway Protocol (IGP)
17	User Datagram Protocol (UDP)

Error detection is provided at the IP level, but the user data is not checked for accuracy, only the IP header. This provides some streamlining of the protocol processing while still ensuring reliability of the routing information. The *header checksum* checks only the header data, which includes the source and destination IP addresses. When checking the IP header, the checksum validates the IP version number and validates that the time-to-live field does not equal zero. It also checks that the IP header is not corrupted and the message length is acceptable.

The *source* and the *destination IP addresses* are used by the routers and gateways within an internet to route the data unit. These addresses remain the same throughout the life of the data unit and are not altered by intermediate networks. We will discuss the format of IP addresses in greater detail in the following section.

An optional *option* field is provided for specific applications but is not always used. The options field is usually used by network control or for debugging purposes. When the *record route* feature is used, for example, the option field will indicate this. The record route data then follows the option field.

The data provided by IP options is variable and will depend on the actual application using it. For this reason, *padding* is provided. Padding is the placement of all zeroes to maintain a 32-bit boundary within a protocol. If the variable data causes the information to be less than a 32-bit segment, the rest of the space is filled by zeroes (as filler) and ignored.

Figure 8.3 shows an entire IP header, including the option field with the *record route* option enabled. The record route data fields are shown after the option field. The data field contains the layer 4 header as well as the customer data. This field may consist of a fragment instead of an entire data unit, depending on the size of the data unit when it was received by layer 4 and depending on the services being provided by the upper layer. IP does not care about the contents of this field and has no visibility to its data.

IP Header

Version	Length	Service Type		Total Length	
Identification				Flags	Fragment Offset
Time to Live		Protocol		Header Checksum	
Source IP Address					
Destination IP Address					
Copy	Class	Number			Padding
Data					

IP Options →

Figure 8.3 IP header with options field.

8.1.2   IP addressing

Addressing occurs at several layers within any protocol. For data to flow from a computer application, through the computer to an external interface, over a LAN, and out through a router to an internetwork, there must be addresses that identify all of the various interfaces used along the way.

The application will create a username. This is the first level of addressing and identifies a specific user to receive data. For example, if you are sending e-mail to someone, you do not know his or her machine address or individual IP address. All you know is the e-mail address, which can be translated by the receiving mail server into a username and machine name.

The internet application will identify a port to be used for the transmission. Consider the port as an internal logical address, not a physical address. A socket is assigned by the operating system and is a combination of the port number and the IP address.

The transport layer will identify an address for the protocol to be used to translate the data and present it to the application. For example, data generated using a TCP application must then be received by TCP at the destination so that it may be processed and presented in its original form to the application.

The network layer will depend on IP addresses to route data units through various networks to reach the destination. Routers will read the IP address to determine which physical port the data unit should be transmitted through. The addresses we have already mentioned are transparent to the IP layer and are processed by host-resident software only.

SAPs are used by the LAN protocol for addressing within the LLC. This is not part of the IP protocol but is part of many LAN protocols such as Ethernet and Token Ring. The LLC is defined in the IEEE standard 802.3.

IP addresses are used at the network layer (layer 3) to route data units through the internet. There are four classes of addresses supported, although the standards are moving away from class addressing to classless addressing. This provides much more flexibility in the addressing scheme and allows for more IP addresses to be assigned.

Figure 8.4 shows the structure of the four classes of addressing. Class A addresses are used most commonly within private "closed" networks. A closed network is one in which there is no external connection. Class B addresses are also used within closed networks. Class C addresses are the most commonly used address and are used for communications to external networks.

Figure 8.4 shows that a Class A address supports 126 network IDs and a total of 16,777,124 hosts. This class of address would not work for large internets because the number of networks outweighs the number of hosts. In a large internet, the nodes within the internet (routers and gateways) do not care about the individual hosts. They only care about the network

IP Address Formats

0 1 2 3 4 5 6 7 8 9 10 11 12 13 14 15 16 17 18 19 20 21 22 23 24 25 26 27 28 29 30 31   Bit Position

Class A 0-127	0	Network Identifier	Host identifier	
Class B 128-191	10	Network Identifier	Host identifier	
Class C 192-223	110	Network Identifier	Host identifier	
Class D 224-239	1110	Multicast Address		
Class E >239	11110	Reserved		

Figure 8.4   IP addresses.

ID. So Class A addresses can only be used within closed networks.

A Class B address supports 16,384 networks and 65,534 hosts. This is designed for medium-sized networks and is not suitable for large internets. A private internet used to interconnect smaller networks within a large corporation would use Class B addresses.

Class C addresses are used for connections to the Internet itself. The Class C address supports 2,097,152 networks and only 254 hosts within a network. This is well suited for connecting to the Internet because it supports a large number of network connections.

Class D addresses are considered multicast addresses. Within a network there can be a multicast address, which is used to reach a group of individual hosts. Each host is assigned a multicast address and can be addressed as a group through the Class D address assignment. Routers manage the multicasting of data units to these addresses and maintain the table of hosts that are assigned these addresses.

The IP address does not identify the physical address of a particular machine. Remember that the actual computer may be connected through a LAN, which requires an address of its own (machine address). The machine address used by LANs is not compatible with IP because of the address structure. The IP address identifies a machine's connection to the IP network.

Consider this scenario. A computer is connected to the LAN using an Ethernet card. The Ethernet card has a machine address hard-coded into the card (nonchangeable), which is used to deliver data units over the LAN. As a new data unit is created, the IP address is embedded into the data unit by the originating host and then passed down to the LLC, which in turn will assign another address for use at the LAN level (the SAP).

The LLC then passes the data unit down to the MAC layer, which will embed the machine address (from NIC) into the data unit and transmit the data unit over the LAN to the router. The router will not send the MAC or LLC addresses over the Internet; it strips them off, leaving only the IP address at the network layer.

This again shows the flexibility offered by layering protocols. The addressing required at each layer can be included without interfering with the other layers. Without layering, this would be a difficult task. If a host is moved within a network to a new connection, the machine address does not change, nor does the IP address. If the machine address changes (because a new NIC card was inserted), the IP address may need to change, but only within the server that provides mapping from IP addresses to machine addresses. The machine can move within a network without requiring the IP address to change. It makes no difference where a computer is as long as it has an IP address to identify its connection to the Internet. Some argue that this is not efficient use of IP addresses since a machine does not stay connected to the network all of the time. Because of this, some networks use dynamic IP addressing.

Dynamic IP addressing requires a machine to log into a server and obtain its IP address from the server upon connection. The server will then provide the next available IP address to the machine. The server is able to keep track of the assignments by machine address because the request for an IP address is encapsulated in a LAN data unit, including the machine address. This is a more efficient use of the IP addresses because they are only used when a connection to the Internet is needed.

There are cases when the IP address does not sufficiently meet the needs of a large corporation. For example, Acme Corporation may need to identify more than one network. It may have a need to address its various offices independently. This would require a Class C assignment for each location, rather than a Class C assignment for the corporation. This is when subnetting is used.

The Class C address identifies the corporation's series of networks. By using a subnet mask, the corporation can further define which of its networks should receive a particular data unit. This is extremely useful in addressing several networks within a Class C assignment and has been extended for use in classless addressing.

8.1.2.1 Sockets and ports. IP addresses and port numbers combined create a unique socket address, maintained and monitored by the operating system. The socket identifies a logical entity above the LLC layer. Remember that ports identify an application and usually have predefined numbers from the IP standard.

A socket is the combination of the originating IP address and the port number. The operating system provides the socket and maintains the logical connection established by the protocol. The socket concept allows multiple users (identified by the IP addresses) to address the same application (identified by the port address). This could be compared with the session established by a user in a mainframe environment. This concept was derived from the University of California, Berkeley version of UNIX back in the 1960s.

It is important that ports be standardized and well defined in the IP standards. This allows host-resident applications to identify ports without conflict. There is room for proprietary ports or ports not identified by the standards. These can be assigned within a private network for use within a corporate internet.

The concept of sockets and ports is an important one to understand. Earlier we discussed the various layers of addressing used within a protocol. Together, these two addresses are used internally (within a computer) to route data units to the proper application.

8.1.2.2 Subnet masking. Gateways only address networks and do not look at the host portion of an IP address. Large corporations may need to address different segments of their network using different network IDs for each segment. This would require Class C assignments for each network segment if there were no mechanism within IP to accommodate this requirement.

Fortunately, IP provides for subnetting of an IP address by using a subnet mask (see Fig. 8.5). This allows an IP address to be divided into various subnets. The subnet address remains transparent to gateways and is only processed by the local router or gateway providing access into the corporate network.

The subnet mask uses portions of the host ID for identifying the subnets. The subnet mask identifies which portions of the host address should be used to identify the subnet. This, of course, will further limit the number of actual hosts that can be addressed within a network, but it allows for the division of larger autonomous networks.

The subnet mask is known only to the local gateways, which provide access to the various subnets. The main gateway providing access to the corporate network looks at the network ID portion of the IP address and determines that the data unit is destined to its own network. It then sends it to a routing table where the subnet mask is determined; then the data unit can be sent to the appropriate router (or gateway) for processing. The subnet router then checks to verify that the subnet address is the same as its own and processes the host portion of the IP address for routing to the appropriate host within the subnet.

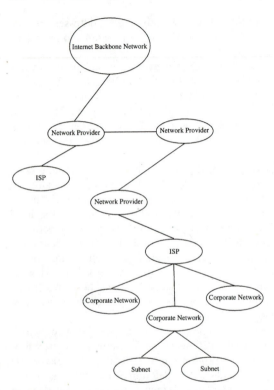

Figure 8.5   Internet subnets.

This is a hierarchical approach to routing within an internet and has many advantages. The biggest one is that intermediate networks used to reach the final destination do not need to know about the various subnets within the corporate network. The corporate network can then be maintained outside the realm of an internet, which allows network administrators to change their addressing schemes within the autonomous network without affecting the rest of their internet. Changes made within the subnets are transparent to external networks, and routing tables in intermediate networks do not have to be updated (again, because subnets are transparent to the rest of the internet).

The routing table within the gateway must be aware of the subnet mask and know how to resolve addressing within its own network. This is a far better approach than assigning individual addresses to every subnet and updating routers and gateways throughout the internet every time these addresses are changed. It also prevents intermediate routers and gateways from having huge routing tables by minimizing the number of addresses of which they need to be aware.

The subnet mask is the same length as the IP address. When converted to binary, the subnet mask usually consists of a string of binary 1s, followed by a string of binary 0s. The subnet mask is then ANDed with the IP address. The end result is the subnet address. For example, we will take the IP address of 199.72.6.100 with a subnet of 255.255.246.0. The router will perform the AND operation against these two numbers. The result would look like the example below:

```
IP Address =
199.72.6.100    11000111   01001000   00000110   01100100
Subnet Mask =
255.255.246.0   11111111   11111111   11110110   00000000
Subnet
Address =
199.72.14       11000111   01001000   00001110
```

The entry within the routers tables would then look something like

Destination	Subnet Mask	Next Hop	Port #
199.72.0.0	255.255.246.0	199.72.14	A

Any combination of 1s and 0s can be used in a subnet mask. In short, a subnet mask can be very simple or very complex, depending on the needs of the individual network. This method of routing to subnets is critical where classless addressing is used. There are no classes of addresses defined in the newer IP standards because masks can be used to determine the network ID and host portions of an IP address.

So far, we have identified how IP addresses work. They are of no value to the LAN, however. The machine address resident on each computer's NIC is the only way to route data units from an internet over the LAN and to the host. Consequently, an IP address must be converted, or *resolved,* into a machine address. A dedicated server performs this function.

8.1.2.3   Domain name system (DNS).   No one likes to use numbers when sending e-mail or when trying to connect to another network. The *domain name system* (DNS) supports the use of a naming convention for addresses rather than IP addresses requiring numbers. The address must be converted at some point to an actual IP address, but this process is transparent to the user.

DNS was developed by the Internet Network Information Center (InterNIC) in 1983 and is administered by an independent company today. All users of the public Internet must apply for a domain name that corresponds to their IP address. Domain names are a popular possession and have become the trendy identity over the Internet.

DNS uses a hierarchical approach, which fits well within the IP addressing scheme. A dedicated server must be used to resolve the domain name into an IP address. This same resolver also provides the machine addresses of all the local hosts (with a direct connection to the server).

The name resolver is a server application, which must then rely on the services of a name server. The name server performs the actual lookup of the domain name and provides the actual IP and machine addresses of the host. The name server and name resolver only know about local hosts and do not concern themselves with addresses outside of their own autonomous networks.

The name server can reference other name servers. This is often the case when an IP address for an entity outside of the local network must be resolved. Name servers are deployed in such a fashion that they know of all other name servers to whom they have a direct connection. Requests may be sent through a number of name servers before they are actually resolved and an IP address provided.

Once resolved, the name server (and the name resolver) can maintain the IP address in cache memory. This eliminates the need to query name servers for domain names used frequently. The domain names do not stay in cache forever and are eventually removed from the cache (determined by a time-to-live parameter). The time-to-live is determined by the network administrator and stored in the name server. This is not the same time-to-live parameter found in the IP.

Name servers perform in one of two modes, *recursive* and *nonrecursive*. When in recursive mode, the name resolver sends a request to the name server. If the name server cannot resolve the address, it sends a query to another name server within its domain. If that name server does not know how to resolve the address, it in turn can send a request to name servers within its domain. The end result is that the address is eventually resolved and the results returned to the name resolver.

In nonrecursive mode, the name resolver sends a request to the name server. If the name server cannot provide resolution, it cannot send a query to other name servers. Instead, it sends an error message along with the address of another name server that can resolve the address. It is then up to the name resolver to send the query to the other name server.

One name server must be labeled as the authoritative name server for each network. This is the name server administered by the local network administrator. The name server is responsible for a defined zone (such as within a corporate network), and the various "subtrees" within the domain (subnets, for example) are defined as zones. For large corporations with several subnets, a name server may be located within each zone, administered by the local network administrator.

The primary name server for a network exchanges databases with secondary name servers within the same network (or outside of the network) to provide redundancy. Each network usually has a primary and a secondary name server. The name servers use UDP (or they can use TCP) to exchange database information. Likewise, queries to the name servers use the UDP or TCP (usually UDP).

The domain itself identifies the user, the subnet (if applicable), and the domain in which it is located. The actual structure of the domain name address uses the following convention:

```
user.subdomain.domain.domain
```

with the subdomain identifying the subnet. A typical domain name address might look like

```
travisruss@aol.com
```

The company name is almost always part of the domain name, unless the e-mail services are being provided through an Internet access provider, in which case the provider's name would be in the domain name (AOL in the above example). The extension identifies the type of organization to which the domain name belongs. Table 8.3 shows the extensions defined.

The domain name system adds a level of user friendliness to the Internet and has been embraced by all of the TCP/IP community. Today, domain names can be seen everywhere, from newspaper articles to television commercials. They provide a means for identifying not only a user within the Internet community but also the organization and type of organization it is.

8.1.3 Routing in an internet

Routing decisions can be based on two criteria: state of the various nodes and links or distance to the destination. When the distance to a destination is the prime criteria for basing routing decisions, delay, throughput, and the ability to reach various gateways and routers along the path are the only determining factors when a route is chosen.

TABLE 8.3 Domain name extensions.

.com	Commercial
.edu	Educational institution
.gov	Government institution
.mil	Military branches
.net	Major network providers
.org	Nonprofit organizations
.int	International organization

When the state of the network is the prime consideration, a number of criteria are used to determine the best route. Link capacity, the number of packets in queue for any particular link in the route, link security requirements, and the number of hops to reach the destination (distance) must be considered.

Hosts within a network receive routing information from routers within the same network and use this information to update their own routing tables, but they do not send their own routing tables to other nodes. Only routers and gateways propagate routing information throughout the network. Hosts use the routing information for a variety of applications, which will be explained below.

8.1.3.1 Source routing.   With source routing, the host determines the route to be used for a particular data unit. Within the IP portion of the data unit, the host will include the IP addresses of all intermediate nodes, which determine the route for the data unit. This is based on information received from adjacent routers regarding routes. Hosts calculate the best route based on distance to reach the destination rather than state of the various nodes within the route. This technique is commonly used in large LANs but is not efficient for an internet. There are too many variables involved with large internets for this method to be efficient.

Nonsource routing is most commonly used in internets. Routing decisions are based on the destination address in the IP header and the routing information maintained by each gateway and router along the route. As each intermediate node receives a data unit, the destination address is examined, and the receiving node determines a route. As the data unit moves from one gateway to the next, routing decisions must be made at each node. This is a better method of routing because it allows intermediate nodes to make decisions based on the dynamics of the network, which may not be known by the originating host.

Within a LAN, the routers do not process the IP portion of the data unit. Only the machine address is used for routing within the LAN. If a data unit is destined for an outside host, the machine address of the gateway or router is used. The router then examines the IP address to determine the route to take to the destination.

8.1.3.2 Time stamping. Time stamping allows each router to store information regarding round-trip delay for each route. This information is gathered using a routing protocol. As the IP header is processed by each gateway/router, the time the data unit was received is inserted (in milliseconds) into the IP header, along with the IP address of the node recording the time. This continues until the data unit reaches its destination. Once the destination is reached, the time stamp information can be extracted and added to the host's routing tables. Round-trip delay can then be calculated for any number of routes. However, round-trip delay is not a perfect science. There are many variables involved when calculating delay. For example, delay may be introduced because of the type of facility or because of a more temporary condition, such as congestion. To be efficient, these would have to be recorded in real time. This is not possible with most routing protocols.

Time stamping uses the IP header and attaches an option field for the recording of the IP addresses and time stamp of each node encountered over a particular route. Time stamping is not highly accurate because each node is running an independent clock. A clock synchronization network, much like those found in telephone companies, is necessary to make this method highly accurate.

An IP, the Network Time Protocol (NTP), is used specifically for clock synchronization and addresses the issue of accuracy to some degree. It allows nodes to exchange time information for synchronization. One node must be designated as the root node and is responsible for obtaining time from a reliable source. A number of sources broadcast time via radio waves for networks.

Once the root node has obtained the accurate time and updated its own clock, it must send the time information to all of the other nodes in its area using NTP. Obviously, the further away the root node is from the receiving node, the less accurate the time information becomes because of delays within the network. For this reason, it is more advantageous to have many root nodes reporting via NTP to gateways and routers within short distances.

8.1.3.3   Circular routing.

Circular routing is a condition any mesh network needs to deal with. When routes suddenly become unavailable, it is highly possible that data units may be sent to a node and returned over the same link to the sending node. This happens because the receiving node thinks that the link is a better route, based on conditions it knows about on other links. The problem is that the adjacent node may not know these conditions, and when the data unit is received again, it will end up sending the data unit back over the same link. This continues until the time-to-live parameter in the IP header expires or until the routing tables are updated by a routing protocol.

Other methods of preventing circular routing are discussed below.

8.1.3.4   Split horizon.

When routing information is received over a particular link, the receiving node updates its routing tables but does not send any routing advertisements from its own routing tables over the same link. For example, say node A is connected to node B. Node A sends routing information concerning a new route over a link to node B. Node B updates its tables and then sends routing information back to node A. The result is that node A may think that there is a path to a destination through node B, when in actuality, the path to the destination from node B is back through node A. If a data unit were to choose this route, it would pass through A to node B, which would in turn send it back to node A (based on its routing information). Node A would then send the data unit back to node B (once again, because of its routing information). This is sometimes referred to as *circular routing.*

To prevent this, a split horizon prevents node B from sending its routing updates to node A. Node A would then select another route for data units rather than node B. The scenario is a bit more complicated, but for brevity sake we will keep the explanation simple.

8.1.3.5   Poison reverse.   With poison reverse, the same principle is used except for one thing. Updates are allowed in the backward direction, but with a weighting factor of infinity. This is interpreted as an unreachable route and remains as such until the next routing update. The problem with poison reverse is that it increases the size of the routing tables, which in turn take more bandwidth when sending routing table updates to other nodes. If bandwidth is a major concern, this method is not recommended.

8.1.3.6   Triggered updates.   Probably the best solution is a triggered update. Routing updates only occur when conditions change. There is no need to send routing information if there has been no change in the routing table. The more routing exchanges that occur, the more the likelihood for circular routing.

When a change does occur (such as a link failing at a particular router), the router sends an update to its adjacent nodes, showing the route with the failed link as unreachable. The receiving nodes in turn update their tables and send routing advertisements to their adjacent nodes. This creates a cascade of data units through the network, and in itself could introduce problems with bandwidth. To counter the problem, random time delays are set at each router so that they do not all send their routing advertisements at the same time.

8.1.4   IP routing protocols

There are a number of protocols that are used within a network and within an internet by routers and gateways. Remember that TCP/IP is capable of rerouting data units around failed nodes, making it self-healing. This requires a level of intelligence within the internet itself.

Routers and gateways must be able to exchange routing information with one another autonomously. Otherwise, a network administrator would have to continuously update the routing tables manually. This would be a formidable task given the size of many of these routing tables and the frequency at which changes occur within an internet. By allowing routers and gateways to exchange routing information with one another, the networks can maintain the status of all of their neighboring nodes and alert other nodes of any status changes. This must follow a hierarchical approach, however, to prevent a flood of broadcast messages throughout the network.

Routing protocols differ depending on the geography of the network. Gateways attached to the backbone of the network do not need to exchange routing information with routers located within an autonomous network. They only need to communicate with adjacent gateways and only concern themselves with network addresses, not host addresses.

There are also two types of routing protocols. *Link state metric* protocols send link and node information so that routers can make routing decisions based on the state of the network. *Distance vector* protocols include information regarding the number of hops a data unit must travel to reach a destination and the amount of delay encountered. Routers then make routing decisions based on the distance and delay factors.

Link state metric protocols make routing decisions based on the status of the communications link. Capacity of the link, delay and throughput requirements, status of the transmission queue, status of the various routes and the number of hops to reach the final destination, and various other link statistics are the parameters used to make routing decisions.

Gateways using distance-vectored protocols periodically send test messages to adjacent gateways to ensure that the routes are still reachable. Should a route test message fail, the gateway must then update its routing tables and share the updates with its adjacent gateways. This is not real-time processing and results in slow updates to routing tables.

Routing protocols allow devices to communicate with other devices within a geographical area. It is not efficient for all devices to communicate with all other devices. For this reason, routers and gateways are placed into groups, and they communicate accordingly.

Routers are placed into areas and communicate with other routers within the same areas. This prevents routing tables from becoming too large. The network administrators are responsible for the deployment of routers and gateways and should always configure their networks in such a fashion.

A gateway provides access to other gateways outside of the area. Gateways then communicate with other gateways, exchanging network addresses and ignoring host addresses. The intent of the gateway is to provide a path from one area (or network) to another.

Gateways and routers in TCP/IP networks are classified four ways:

1. Interior gateway

2. Border gateway

3. Exterior gateway

4. Gateway-to-gateway

For each of these classifications, there is also a routing protocol. This allows gateways to exchange routing information with other gateways at the same level.

Interior gateways exchange routing information within an autonomous system with other gateways or routers using the *Interior Gateway Protocol* (IGP). These are locally maintained routers that provide access to other subnets within a network or to an exterior network through a connection to an exterior gateway.

The *Exterior Gateway Protocol* (EGP) exchanges routing tables with other routers and gateways connecting other networks. For example, company XYZ may have an exterior gateway connecting it to the Internet. Its gateway would exchange routing information with gateways from other networks (with which it has direct connections) but would not share routing information about destinations within its own network. The EGP may be replaced with the *Border Gateway Protocol* (BGP).

This protocol was first deployed in 1989. However, some feel this is a more robust protocol, and if it is embraced throughout the Internet community, it may indeed replace the EGP.

For gateways outside of any network (part of the internet backbone) the *Gateway-to-Gateway Protocol* (GGP) is used. These gateways are used by service providers to connect autonomous networks to the backbone of the Internet or to a smaller private internet. They exchange routing information about how to route from one network to the next.

It is important to remember the scope of these routing tables. A gateway has information about how to reach all destinations within an internet. The destinations are not individual nodes, but individual networks. Routers and gateways associate routes with physical ports, so it is likely that many destinations will use the same port.

The only portion of the IP address these nodes are concerned with is the network ID portion of the Class C address. It seems that routing tables would be enormous because of the number of users on the Internet itself, but in actuality, routers only need to know about the networks themselves and not the nodes within those networks. Exterior and border gateways maintain information about interior gateways and routers, and these routers in turn maintain routing information about the individual nodes within their own networks.

Also remember that routers and gateways do not communicate with every other router and gateway within an internet. They only exchange routing information with routers and gateways to which they have a direct connection. This helps prevent a flood of messages from congesting the network. It also provides a check and balance mechanism with which routers and gateways can keep track of the status of neighboring nodes.

All routing protocols use TCP to exchange routing information. The frequency at which this information is exchanged depends on the protocol used and the configuration of the network itself. It may also depend on network activity. If activity is high, the routers and gateways will not communicate as frequently as when the activity is low.

8.1.4.1 Address Resolution Protocol (ARP).

The *Address Resolution Protocol* (ARP) is part of the IP stack and is an IGP. ARP provides translation of IP addresses into machine addresses. This requires communications with end nodes or computers on a LAN.

The router within a LAN uses mapping tables, which map the IP addresses received to the proper machine address. Of course, before this can be done, the router must know the machine address. The router communicates with all nodes on the LAN to determine which machine addresses are reachable and what their IP addresses are. When the router receives a data unit, it checks its mapping table to see if there is an entry for the IP address received. If there is, the router can create a packet (such as an Ethernet packet) and transmit the data over the LAN.

If there is no entry for the IP address, the router sends an ARP broadcast message over the LAN. All active nodes on the LAN will see the broadcast message and process it. If a node recognizes the IP address in the broadcast message as its own, it will reply with its machine address. This is then placed in random access memory (RAM) cache in the router for temporary storage.

The cache entry is not a permanent entry. There is a parameter associated with every entry that determines how long the particular entry remains in cache. When the entry expires, the router has to send a broadcast message again to receive that node's machine address.

The purpose of the ARP cache is not to provide a permanent record of all hosts on a LAN (because these hosts are not always on and accessible) but to provide a temporary mapping table so that the router does not have to continuously send broadcast messages over the LAN. The ARP cache can also identify the hardware type (such as Ethernet or Token Ring), which is important for multiprotocol routers. These routers may have connections to several different types of LANs, and they need to know what type of protocol data unit needs to be created to route the IP packet to the host. Also part of this table is the type of protocol used to obtain the routing information (such as ARP or RIP), the routing age (how long the information has been in cache), and the subnet mask for the destination host.

Routers use ARP when addressing hosts that know their IP address. This requires a workstation (or PC) with some form of permanent storage (such as a hard drive). The IP address must be configured into the host (through IP client software) so that the PC knows its identity. Not all workstations know their IP address because they may not have disks, in which case a different protocol is required.

8.1.4.2 Reverse Address Resolution Protocol (RARP). Some workstations may not have hard drives (diskless workstations) and are not capable of storing their IP configuration. In such cases, the workstation must send a query to the server to determine what its IP address is. The *Reverse Address Resolution Protocol* (RARP) allows diskless workstations to send queries over the network (in broadcast mode). Servers on the network read the machine address from the query and send a reply with the workstation's IP address.

8.1.4.3 Routing Information Protocol (RIP).   The *Routing Information Protocol* (RIP) is an interior gateway protocol used by routers to exchange routing information within an autonomous network. RIP is based on two Xerox protocols, PUP and XNS.

RIP is a distance-vectored protocol. It is the most popular of the routing protocols. The routes are determined based on distance, without consideration of the state of the links and/or nodes that must be used to reach a destination.

Two modes of communications are used with RIP, passive and active. Passive mode is used by workstations, which read RIP messages and update their routing tables but do not exchange routing information with other nodes. In active mode, devices read and send routing information to adjacent nodes. Routers and gateways are the only devices that operate in active mode. When a router or gateway receives an RIP message, it updates its own routing table and then generates its own RIP message to send the updates to its adjacent routers/gateways. If no updates are received for a given route, the route is marked unreachable. The assumption is made that the route has either failed or is unreachable and should therefore be removed from the routing choices.

RIP is not an efficient routing protocol because it sends routing advertisements on a periodical basis rather than on a select one. Consequently, the network is burdened with extra traffic that may not be necessary.

8.1.4.4 Open Shortest Path First (OSPF).   *Open Shortest Path First* (OSPF) is a link state metric protocol. It makes routing decisions based on link capacity, delay and throughput requirements, the number of data units presently in queue for transmission over a particular link, the number of hops required to reach a destination, and the ability to reach gateways and routers along the route.

This makes OSPF more robust than RIP, which bases routing decisions on the number of hops to reach a destination. In RIP, there is no consideration of capacity or link state. In OSPF routing, tables are created based on the above criteria, and a weighting factor is applied to each route. This allows for routing based on the dynamics of the network. It also allows routing decisions to be made based on the type of facility. For example, a satellite link may be given a higher weighting factor even though it is the shortest path to a destination. Satellite links introduce additional delay, and even though there may be fewer hops through a satellite link, there could be longer delays by taking such a route.

Another reason for the popularity of OSPF over RIP is the frequency at which routing advertisements are sent. With OSPF, these advertisements can be made more selectively, rather than being based on time. The only time advertisements need to be made is when changes to existing routes are made (such as the case when a router or link should fail). The advertisements are then sent in response to a condition rather than in response to a clock.

OSPF is not one protocol but a series of routing protocols consisting of test messages and routing exchange messages. It allows routing based on the dynamic state of the network rather than distance.

8.1.4.5 HELLO protocol

HELLO is used by OSPF for communicating with adjacent nodes. This protocol first establishes an adjacency with another node and then continuously monitors that node (through test messages) to ensure it is still reachable.

An adjacent node can be one of several states. When a node is first placed onto the network, it is considered to be in the DOWN state. This indicates no adjacency has been established. Once a HELLO packet has been received, the node is placed in the INIT state by the sending router.

When a response is sent to the HELLO packet, the node is moved to the two-way, or EXSTART, state, which implies that communications have been established, but adjacency is still being negotiated. Routing information cannot be exchanged until the node has reached a FULL state. Once adjacency has been established, the two routers can begin exchanging routing information.

Rather than have all routers within an autonomous network share their routing information with a gateway, one router can be designated as the one to exchange routing tables with a local gateway. This prevents routers from creating congestion by advertising their routes to the same gateway.

8.1.5 IP services

IP provides many services other than routing (although these other services are associated with routing of data units). Like other protocols, IP must manage all aspects of the data transmission over the network. Keep in mind that IP is classified as a network layer protocol. This means that IP must provide reliable transfer of data from one network to another through the use of routers and gateways.

8.1.5.1 Fragmentation and reassembly. Fragmentation is the division of data units into smaller data units. This is an important feature in any network and can be a complex issue as well. IP provides fragmentation across networks, which means a data unit can be fragmented into several data units at the source, and then the various fragments can be fragmented even further by intermediate networks.

Two fields within the IP header support fragmentation, the "more" bit and the fragmentation offset. The more bit indicates that there are more fragments to be received, and the receiver should not reassemble the data unit until the last fragment is received. The receiver uses the fragmentation offset to reassemble the data unit. It identifies where the fragment belongs in the entire data unit. The offset is always in relation to the original data unit as it existed at the source, not after fragmentation by an intermediate network.

It is possible that a fragment could be received by an intermediate network and be fragmented further because the received data unit is too large to pass through the intermediate network. In fact, data unit fragments can take different paths and arrive out of order. IP ensures that all of the fragments are received and can be reassembled properly.

To ensure that all fragments have been received, IP uses a timer. All fragments must be received before the timer expires, or an error message will be returned to the originator, resulting in a retransmission of all of the fragments. IP cannot retransmit one fragment of a data unit because there is no way for IP to determine which fragment was lost. In the event a fragment is lost, all of the fragments are retransmitted, and received fragments are thrown away.

8.1.5.2 Error control.   IP has no error control and relies on the ICMP for the delivery of error messages. ICMP resides in hosts and gateways, and it provides administrative and status messages. Routers cannot generate ICMP messages. Typically, gateways generate an ICMP message with the originating host as the recipient. This means that the ICMP software residing in gateways is more complex than that found in hosts. Since gateways are usually the message generators, there is no need for complexity in the host, which saves in memory and processing.

An ICMP message is generated if a destination is deemed unreachable by a gateway, if the time-to-live field expires for a data unit, or if a gateway determines that a header is in error. If an error occurs with an ICMP message, no report is generated.

ICMP reports errors with fragments, but it only reports the first fragment in error, not all subsequent fragments. The information field of the ICMP message contains the first 64 bits of the fragment in error, so the host can determine how to handle the error.

It is important to understand that ICMP does not provide error detection for IP. It is simply a reporting mechanism used by IP to report errors to originating hosts. IP is still a connectionless transport, which means delivery is unreliable and not guaranteed. IP relies on upper-layer protocols (such as TCP) to provide robust error detection and correction. ICMP is considered part of the IP suite, but it uses the services of IP for delivery of error messages. It is carried in the data portion of the IP header. The IP header field "protocol" indicates the message type as ICMP (see Fig. 8.6).

Within the ICMP header is a type field, which indicates why the ICMP message was generated (for instance, "destination unreachable"). There are 13 types defined in the protocol. The code field provides additional information about the error.

As mentioned earlier, the first 64 bits of the data unit are provided in the information field of the ICMP header. This allows the originating host to correlate the data unit with one already in its transmit queue waiting for acknowledgment.

Information (variable)	Parameters (optional)	Checksum	Code	Type	IP Hdr

Figure 8.6   ICMP header within IP header.

8.2   TRANSPORT CONTROL PROTOCOL (TCP)

TCP is a layer 4 protocol and is a user of IP services. It does not require IP, however, and can use almost any other layer 3 protocol for transport services. The software resides in hosts but not in routers. A router does not need to know about TCP because it does not process the TCP header. Only hosts (both originating and receiving) process this header.

There is one exception to this rule. Sometimes routers will use TCP software for network management purposes, in which case they must be able to generate a TCP header to encapsulate the network management message. User data is never processed in routers. Likewise, gateways do not use TCP (other than for the generation and processing of network management messages). We will discuss network management later on in this chapter, but for now understand that network management is considered an application, and it relies on TCP and IP for the delivery of network management data units.

TCP provides end-to-end session control between two hosts. Remember that TCP is connection-oriented, which requires a session to be established before data can be exchanged. TCP also provides error detection and correction for applications requiring connection-oriented services.

As with any connection-oriented protocol, once the session is established between two logical entities, the receiving host must acknowledge all received data units. If a data unit is not acknowledged, no error messages (negative acknowledgments) are sent by TCP. Instead, the sending host sets a timer, and if no positive acknowledgment is received before the expiration of the timer, the data unit is automatically retransmitted. This works independently of any IP processes.

Before discussing more TCP services, we need to examine the TCP header and understand the various fields in it. The TCP header is in the data portion of the IP header.

8.2.1   TCP header

The data units that are sent between two entities are called a *segment*. The segment is sent down to the IP stack where it is encapsulated by an IP header (see Fig. 8.7) and becomes a *packet* (recall the discussion at the beginning of this chapter).

The addresses contained in TCP are somewhat different from the IP addresses we discussed in the previous section. Addressing at the TCP level is of logical entities within a host rather than actual user connections to the network. Recall that the IP address is not a physical address; it indicates a connection to the network and identifies a user.

TCP uses a destination and source port number. A port is a predefined number identifying the application using the TCP services. This may be FTP, TELNET, or Simple Mail Transfer Protocol (SMTP or e-mail). The port number is 16 bits long. A 32-bit sequence number is used to verify that all data units have been received. This is a number sent in serial fashion with every data unit. Sequence numbers can be received out of order as long as they are received within a time constraint (determined by a TCP timer). If a sequence number is received out of order and the timer expires, all unacknowledged sequences must be retransmitted.

Source Port (16 bits)							Destination Port (16 bits)	
Sequence Number (32 bits)								
Acknowledgment Number (32 bits)								
Data Offset (4 bits)	Reserved (6 bits)	U R G	A C K	P S H	R S T	S Y N	F I N	Window (16 bits)
Checksum (16 bits)							Urgent Pointer (16 bits)	
Options (variable)								Padding
Data (variable)								

Figure 8.7 TCP header.

An acknowledgment number follows the sequence number and identifies the next expected sequence number. The acknowledgment number is 32 bits long. There are no negative acknowledgments, only positive ones. This field is considered a positive acknowledgment. It identifies which sequences have been received properly and which sequence number is expected next.

If the originating host has already sent the sequence number identified in the acknowledgment number field, the host will not retransmit until a timer expires. The originating host allows a certain amount of time, which is determined by calculating the round-trip time for the route used, before retransmitting the sequence again. Since this is a dynamic value, the process can be somewhat complex.

The data offset field identifies where the data portion of the TCP header begins. It identifies how many 32-bit words are in the header, preceding the user data field. This field is 4 bits long and is necessary only because there is an optional field at the end of the header, which can be of variable length.

Several 1-bit fields follow the data offset field and are used for processing the TCP data unit. The urgent bit identifies that the urgent pointer contains data. The urgent pointer is a 16-bit field that identifies the offset in the user data field that contains urgent data.

The acknowledgment bit identifies that an acknowledgment is present in the acknowledgment number field, and it alerts the receiver that this number is acknowledging previously received sequences. The receiving host knows then to delete the sequence numbers being acknowledged from its transmit queue.

The push bit is similar to the urgent bit. It notifies the receiving host that the data unit received should be processed immediately and causes the data unit to be processed when received rather than being placed in a receive queue.

The reset bit causes a session (logical connection) to be reset. This usually means that all queues associated with the session are flushed and all associated counters and timers are reset to zero. This is used when an error occurs with a connection and the connection must be reestablished.

The synchronized bit is used when establishing a logical connection, and it indicates that sequence numbers need to be synchronized. Synchronized sequence numbers allow both hosts to identify where the sequence numbers will begin so that each entity knows what sequence numbers to expect. This bit is used during the handshaking process, which takes place during the connection establishment phase.

The finish bit is the same as end of transmission. It indicates that there is no more data to be sent, and a session can be closed. The session will then be terminated and the resources released for another session.

The window field is used during session establishment. Each host must negotiate how many data units can be sent before an acknowledgment. This is considered the window size and is determined by the size of the queue and the amount of processing already occurring from other sessions. The window size cannot be changed once the session has been established.

The checksum field is used to check for errors in the header as well as in the user data. This is different from IP, where it only verifies the header itself. The checksum in the TCP header does verify the user data.

The options field is a variable field designed for future TCP implementation. There have been no uses for this field at the time of this publication. Padding follows the options field to maintain a 32-bit boundary.

8.2.1.1 Processing of urgent data.
Urgent data can be an interrupt, control information sent from a terminal, or almost any other data requiring real-time processing (or near real time, because TCP/IP is not a real-time protocol). The method used in TCP to identify urgent data is sometimes referred to as *out-of-band notification,* which means that data must be sent outside of the data stream. For example, if a terminal emulation application (such as TELNET) were running, the normal data stream would contain data being transferred to and from the remote terminal. Urgent data (control characters on the keyboard) is processed outside of this data since it is not part of the information being sent or received.

When a host receives data marked urgent, it processes the data identified by the header (the urgent pointer field). The data is not placed in a buffer, as the normal data stream would be. This ensures that the data is processed when received without delay. Urgent data will also receive a sequence number, but because this data is processed immediately, it leaves a gap in the sequence numbers received in the buffer. Software must manage this gap so that when the data in the receive buffer is processed, the host knows that the missing sequence numbers have already been processed.

The urgent pointer in the TCP header identifies where the urgent data ends rather than where it begins. This allows other normal data to be received in the same data unit as urgent data. The host processes all data up to the pointer and then leaves the urgent mode and begins its normal data processing. If the window size has been reduced to zero, urgent data is still processed. Again, this ensures that all data has been processed and is not left in a queue or ignored because of processing capacity.

8.2.1.2 Processing of push data. The push bit also provides a means for bypassing the receive buffers. Push data can be buffered if the data unit is fragmented. This is different from urgent data, which is passed to the application without buffering. The difference is that urgent data is normally small data units and does not require fragmentation, whereas push data may be larger.

When multiple data units are used for push data, the data stream is buffered until all of the other fragments have been received. If all of the data is not received within a specified time frame (managed by a TCP timer), the data already received is passed to the application. If there are missing fragments, the application may return an error, indicating that not all data has been received. It is the responsibility of the application, rather than TCP, to report this.

8.2.2 TCP ports and sockets

We discussed ports earlier and identified a port as a predefined number used by the protocol to indicate the application using the data unit. This tells the host which upper-layer application to pass the data unit to.

A socket is something managed by the operating system and is made up of both the port number and the originating IP address. Together, the port number and the source IP address form a socket identifier, which is used to identify logical connections established by TCP.

All sockets must be unique since a pair of sockets (originator and receiver) identifies a session. The socket is of significance only within the host. The originating host will know the socket number but will not manage the resources required to support the socket at the destination host. Each host is responsible for managing resources allocated to each socket within its own entity.

The defined port numbers commonly defined by the TCP protocol are

- 20 = FTP data

- 21 = FTP control

- 23 = TELNET

- 25 = SMTP

- 42 = Nameserv (host nameserver)

- 53 = Domain (domain name server)

- 109 = Post Office Protocol (POP) 2 (used by e-mail applications)

Many other port numbers have been defined, but these are the most common. The port numbers 0 through 255 are predefined (cannot be defined by network operators), whereas any port number above 255 can be defined by network operators.

A port is capable of supporting multiple sessions. The operating system manages the resources needed to support multiple users. The actual number of sessions that can be supported depends on the platform and the operating system.

8.2.3   TCP services

TCP provides a number of services to hosts. Remember that TCP is a connection-oriented protocol and must provide session management as well as reliable transfer of data units. IP is a connectionless protocol and relies on TCP to provide reliable data transfer.

In this section we will discuss two important services provided by TCP, error control and flow control. These are typical in any connection-oriented protocol, but TCP uses mechanisms that are somewhat unique.

8.2.3.1   TCP error and flow control.   If a sequence is received in error (detected by the checksum), the data unit is deleted from the buffer. No reply is returned to the originating host by TCP. IP checks only the checksum of the IP header, not that of the user data (which includes the TCP header and the user data).

The originating host sets an acknowledgment timer when data units are sent. In the case of an error (either checksum or other) the timer will expire and the originating host will retransmit the data unit. The same applies to data units received out of sequence.

The timer is not a fixed timer, but a variable one based on round-trip time. Round-trip time includes delay in the forward direction (or time to reach destination), processing time at the destination, and delay time in the backward direction (or time for acknowledgments to reach the originating host). This is not a very accurate mechanism because round-trip time is very difficult to determine. There are many variables in the network, including network congestion and node failures, that affect round-trip time.

TCP uses both acknowledgments and windowing for flow control. The TCP header contains fields for both the acknowledgment number and the window size. The actual window (which is the number of data units that can be received without acknowledgment) is determined by both the acknowledgments outstanding and the window size value in the TCP header. For example, if the acknowledgment number sent is 6 and the window value is 12, the actual window size is 18 (acknowledgment number + window value). This is more efficient than using positive acknowledgments only because it allows the host to adjust its window based on resources available and on buffer size. The window size is negotiated when a session is established by TCP. Once the window size has been established, the value cannot be changed by the host without negotiation. This prevents the host from arbitrarily changing its window size at any time.

TCP also maintains an internal window, referred to as the congestion window. This window is not advertised to other hosts. It allows for the throttling of data units transmitted based on retransmissions. When the acknowledgment timer maintained by the originating host expires, the originating host assumes that the destination is congested and enters into the congestion mode.

While in congestion mode, the congestion window is decreased. The value is derived by taking the window size sent in the TCP header and the number of retransmissions sent. If retransmission continues to occur, the congestion window continues to decrease using what is called multiplicative decrease. This means that the congestion window is decreased using a multiplicative formula, with the number of retransmissions being the factor. When there are no retransmissions being sent, the originating host assumes there is no congestion at the destination, and the congestion window will equal the same value as the window size advertised in the TCP header.

When congestion subsides, the congestion window does not immediately change to the window size value. There is a slow restart implemented so that the destination is not flooded with data units, causing it to become congested again. During the restoration process, the congestion window steadily increases until it reaches the same value as the advertised window size. If a retransmission to the congested destination occurs during the restoration process, congestion mode is entered again, and the congestion window is decreased using the multiplicative decrease algorithm.

8.2.3.2 TCP management. Besides error detection/correction and flow control, TCP is also responsible for maintaining established sessions. Software resident in each host is responsible for establishing and maintaining sessions. Remember that a session is not a physical connection between two entities, but a logical one controlled by software (both TCP and the operating system).

TCP keeps information about every session in what is referred to as a *transmission control block* (TCB). The information kept by the TCB includes the source and destination addresses, ports assigned to the session, round-trip time (calculated by the individual host), and data units (sequences) sent and received. In addition, the TCB at the originating host maintains records on acknowledgments and retransmissions sent to the destination as well as statistics gathered regarding the connection (such as the number of retransmissions and number of bytes transmitted).

This data is maintained for every session established by a host. There may be several sessions to one host (especially in the case of a server), which means TCP must maintain many sessions at one time. Some operating systems are better at maintaining multiple sessions than others (such as UNIX). This is why some platforms may be better for TCP/IP applications than others.

8.3   USER DATAGRAM PROTOCOL (UDP)

UDP is called a connectionless protocol, but it really is not. The operating system maintains information about each active UDP socket, which implies a connection-oriented service. In a true connectionless service (such as IP) there are no sockets maintained. Data units are sent to a specific destination, with a port address. The socket is not needed because the data unit is processed when it is received by the application, and there is no further action to take with the originating host.

What UDP does not provide is error correction and flow control. There are no acknowledgments sent for received UDP data units; they are assumed received. The application using UDP services is responsible for determining if there have been errors or if data units are missing. This makes UDP an unreliable service, but reliability is not always a concern with applications such as e-mail and some network management functions. Upper-layer protocols can make up for what UDP is lacking. This provides a streamlined protocol that does not require a lot of processing at the originating or destination hosts. If a lot of processing is not required, data units can be sent and received with very little delay. Many newer protocols operating at this layer (layer 4) and the lower layers use this philosophy.

8.3.1   UDP header

The UDP header is much simpler than that used in TCP (see Fig. 8.8). There are no sequence numbers, which means no acknowledgments. Likewise, there is no acknowledgment number field or urgent data processing capability. There is a source and destination port number and length and checksum fields. Nothing else is needed. Obviously, this allows receiving hosts to process UDP data units rather quickly because all that is needed is to send the received data unit to the proper application (identified by the port number).

8.4   TELNET

TELNET is an application protocol that relies on the services of TCP/IP for delivery. This is also considered a peer-to-peer protocol, residing only in the origination and destination hosts.

The TELNET protocol allows terminals or PCs acting as terminals (terminal emulation) to communicate with remote hosts. The remote host may be a server running an application or a mainframe. Terminals are proprietary and are usually not compatible with one another. DEC terminals use a different terminal protocol than IBM terminals do. These differences must

16 bits	16 bits
UDP Source Format	UDP Destination Port
UDP Message Length	UDP Checksum
Data	

Figure 8.8   UDP header.

be resolved when connections are established with remote hosts. This is the task of TELNET. This protocol provides a means for all terminal types to communicate with remote hosts regardless of the terminal type. It allows the host to determine the characteristics of various terminals.

This is done using a series of commands to negotiate services the host can support. The host must agree to each of the services before the session can begin. All commands are sent using a format defined by TELNET. One example of services to be provided by the remote host is echo, where all data is echoed by the remote host back to the terminal. Other services include control character mapping (defining which keyboard keys will emulate specific control characters) and how flow control is to be managed.

The terminal characteristics supported by the TELNET protocol depend on the terminal emulation being used. However, the main purpose of TELNET is to allow a mechanism by which the remote host can determine what terminal characteristics will need to be supported.

There are many uses for TELNET, and it has many advantages. Without TELNET, terminals rely on modem communications or dedicated leased lines to reach remote hosts. With a connection to an internet, virtually any remote host running the TELNET client software can provide remote access services to all hosts on the same internet.

Corporate users of the public Internet have found it very cost effective to connect their corporate servers to the Internet, allowing employees in remote locations to access the company servers without a dedicated connection. This means they can connect from anywhere that they can get a connection to the public Internet.

Once a connection has been established, TELNET operates as any other terminal emulator. What the users see on their screen is the data provided by the remote host, in the format supported by the remote host.

8.5   FILE TRANSFER PROTOCOL (FTP)

FTP is used to establish a session with remote hosts to download (or upload) files. Downloading files means transferring a file resident on a remote host to the local computer. Uploading files means transferring files from a local computer to a remote host.

FTP uses some of the services of TELNET to establish a session. TELNET services are needed to allow users to view directories on remote hosts and navigate around the host to find files. However, users cannot run applications on remote hosts using FTP.

Two sessions are needed to run FTP. One session is used for sending control information (such as TELNET commands). The other session is used for the actual transfer of data files. This allows files to be transferred without interruption, while still allowing control information to be sent to the remote host.

FTP supports several data types. Files can be transferred in ASCII or EBCDIC formats. These file types are used when transferring text files. Files from word processors and desktop publishing applications cannot be sent using these formats because they include proprietary characters not supported by ASCII or EBCDIC. If files of this nature are transferred using ASCII or EBCDIC, they will be unintelligible because the formatting characters are only recognized by the application that created them. All formatting characters will be translated into ASCII text, which will result in a lot of garbage.

For all other file types (including word-processed or desk-top publishing files) FTP supports binary file types as well. When transferred as a binary file, the file is not converted at all but is sent as is. Binary type is used for all files other than plain text files; this includes image files.

FTP uses the services of TCP and supports authentication and security functions. Some remote hosts may require a login and password (supported by TELNET), whereas others allow an anonymous login, which is accessible by anyone.

8.5.1   Trivial File Transfer Protocol (TFTP)

Trivial File Transfer Protocol (TFTP) is less complex than FTP. It does not support any authentication or security features, which means it does not need all of the TELNET features. It does still rely on TELNET for establishing a connection and negotiating the session.

TFTP does not require the same level of services that FTP does, so it can use UDP rather than TCP. This means that TFTP is not as reliable as FTP because UDP does not provide acknowledgments or error control of any type. For this reason, many network operators do not use TFTP.

When transferring data, TFTP sends a block of data using UDP and then waits for an acknowledgment from TFTP at the distant end. Once an acknowledgment is received, another block of data can be sent. This increases the amount of processing to some degree. TFTP also provides some limited error handling.

Sessions are tracked using a random source transfer ID. The destination transfer ID is always 69, which is a well-known port identifier for TFTP. The source transfer identifier is unique for every session established.

Five different data unit types are used by TFTP. A read request asks to read a file from the destination directory. This is used to transfer a file from the selected directory to the remote directory. A write request is used when modifying a destination directory. Modifications include deleting (provided the operating system provides the necessary user permissions), renaming, or moving files to other directories on the same host. A data type is used for the actual transfer of blocks of data. An acknowledgment type is used to acknowledge receipt of data and is sent after each block of data. If an error occurs, an error type is used to indicate the type of error that occurred.

FTP is a very popular application within the Internet. Many hosts supporting FTP capability are loaded with files that can be downloaded at no cost. The server must have FTP host software, and the remote host must have client software supporting the access of FTP servers.

8.6   SIMPLE MAIL TRANSPORT PROTOCOL (SMTP)

Electronic mail uses the SMTP to transfer mail from one server to the next. Once the mail message has arrived at the destination mail server, the Point-of-Presence (POP) protocol is used to transfer it to the desktop. This protocol uses a spooling methodology, where a data unit is sent to a mail server, which then holds the data unit until the user accesses the server using an SMTP client application. The data unit is then downloaded to the user.

POP supports remote access of e-mail. It stores the mail message until the remote user accesses the mail server and requests mail delivery. Desktop hosts use POP, rather than fully functional SMTP, to access mail servers. With POP, desktop hosts do not need to provide the processing resources required by SMTP. POP is not as complex as SMTP and is only used by hosts accessing mail servers.

Mail servers can be configured to either save a data unit or discard it when it is downloaded. This of course oversimplifies the whole process for sake of brevity. The mail server tries periodically to send the mail to the user, usually until a counter expires, at which time the mail is either discarded or saved on disk.

If the mail cannot be delivered within a certain time, the server can be configured to send an error message back to the sender. This is a function of SMTP and not the underlying TCP or IP protocols. SMTP provides its own error control at the application level.

Clients can be connected continuously to the mail server or can connect intermittently (as is usually the case). When users connect intermittently, it is up to the server to maintain the mail until the user actually connects, and then it can download the mail to the user upon request.

When sending mail to a mail server, SMTP first establishes a connection with the mail server. It then waits until it receives an acknowledgment from the server to send a receipt message. The receipt message identifies the recipient of the upcoming mail. The server can then acknowledge that the username identified in the receipt message is valid (by sending an acknowledgment), or it can return an error message indicating the username is not valid.

When an acknowledgment is received, SMTP begins sending the actual data (mail message). When the transmission is complete, SMTP sends an end-of-transmission followed by a QUIT command. This terminates the session and closes the

connection. Addresses used in SMTP utilize the DNS convention. If the user is not connected to a network using SMTP,
the actual address may follow a different convention than standard DNS addresses. This is because the address must contain
additional information about the non-SMTP hosts used to gain
access to the Internet. For example, the address user%remote-
host@gateway-host indicates that the user is located outside of
the SMTP environment. The standard convention would be
more like user@host.com.

Some e-mail applications support Multipurpose Internet
Mail Extensions (MIME), which allows nontext information
such as pictures, movies, and audio to be attached to the e-mail
as binary files. The files can then be saved on disk for viewing
with a specialized application that is capable of viewing such
file types.

8.7   NETWORK NEWS TRANSPORT PROTOCOL (NNTP)

NNTP is a unique way of supporting special interest groups
without requiring a dial-up connection to some central server.
The newsgroups are similar to bulletin board systems with one
exception: Individuals can connect to all of the newsgroups and
view all of the postings on multiple newsgroups rather than
connecting to one server and then having to connect to another.
By using NNTP, special interest groups are accessible to everyone on the Internet, and postings can be distributed throughout
the Internet.

The way it works is actually simpler than it may sound. A
large mainframe (owned and operated by a third party) is used
to poll news servers from around the world. These "specialized"

servers collect "postings" from local connections and forward them to the mainframe. While sending postings, the news servers also download any new postings that may be in the mainframe computer. However, news servers only download postings for newsgroups to which they have subscribed.

Subscription to a newsgroup is free; all it takes is configuring the server to query the main computer for all postings within that newsgroup, which is an autonomous function. The local server then stores those postings for its subscribers. If the news server were owned by America Online, all AOL customers would connect to the AOL news server, which gets its postings from the mainframe.

The subscribers can only receive postings from newsgroups to which their service subscribes, which allows service providers to pick and choose the newsgroups to which it wants to provide access. This is important to content providers such as CompuServe and AOL, which offer their own versions of competing newsgroups. This is also important to corporations for controlling what their employees can access.

Two students from Duke University developed NNTP so that they could share research notes with other students from North Carolina State University. The protocol uses a store and forward methodology, much like SMTP.

Host software resident on servers collects "postings" and forwards them to a central server, which collects these postings and then forwards them to other servers upon request. This is what makes the Usenet community on the Internet work today. Directories are configured on the various servers, and the postings are placed in the directories from which they were generated. The postings are simple text messages; they can be image files as well. The various servers on the Internet that download postings and upload postings are called *news servers*.

News servers each subscribe to certain newsgroups (or directories), which are topical groupings of postings. On a periodic basis, the various news servers connect to a central news server and download postings from the various groups to which they subscribe. The local news server can subscribe to all of the newsgroups on the news server or to selected newsgroups. The network administrators choose which newsgroups they subscribe to. Users who connect to the local news server can only download postings from newsgroups that their local server subscribes to. This allows corporations to control which newsgroups their employees contribute to.

The NNTP protocol uses the services of TCP. It is capable of sending text or images, as well as other binary files (usually as attachments). The binary files are converted to ASCII text by a utility program so they can be transferred by NNTP. When users receive the files, they must use a conversion program to convert them back to binary files. NNTP does not support the transfer of binary files in their native form. One of the most popular encoding/decoding utilities is uuencode, provided by the UNIX operating system and supported on almost every computer platform.

8.8 HYPERTEXT TRANSPORT PROTOCOL (HTTP)

The WWW has changed advertising from printed media to interactive multimedia. The combination of text with video, animation, and photographs has created an exciting new way to distribute information about products, companies, or almost any form of information. Specialized browsers decode the files, created using Hypertext Markup Language (HTML) or Java, and display them according to rules defined in the browser application. Browsers are applications resident on user computers.

The Hypertext Transport Protocol (HTTP) supports the transfer of files stored on dedicated servers. These files are unlike text or data files. The files are coded using HTML, which is derived from Standardized Generic Markup Language (SGML). Text files are coded with special characters that identify which characters are headings and which characters are underlined, and so forth.

Images can be referenced as well, without being a part of the file. A reference is inserted into the file to point to the location of another file. When the file is opened by a special-purpose viewer application, the user sees the text displayed according to the coding and the images referenced in the source file.

Images and text files do not have to be collocated. In fact, text and image files can be located anywhere on the Internet and linked to the main file by references in the text. This allows files to be maintained locally rather than downloaded to a central server somewhere that must be maintained by a central authority.

To access an HTTP server, a uniform resource locator (URL) is accessed (part of the client), and an address is typed in. The address uses a format that is similar to that of e-mail. The protocol must be identified first (such as HTTP or FTP), followed by a colon, and then the address. The address is in the DNS format, and may look something like

```
http://www.tekelec.com
```

The address varies depending on where the files are located. Most companies using HTTP servers to distribute information use this format with their own company identity, making it simpler to find them even if their address is not advertised (an educated guess should get you to the proper HTTP server).

The browsers have the ability to access mail servers to retrieve e-mail, access news servers to retrieve newsgroup postings, and even access FTP servers to download files from remote computers. All of the functions of the Internet can now be provided through one application program, the WWW browser.

Companies are now looking at these sophisticated browsers for their own internal networks, to view documents stored on corporate servers. This allows companies to distribute all forms of documentation electronically, without the cost of printed copies and without the worry of version control. With only one master file located on the corporate server, changes and updates are guaranteed to get distributed because all who have access will be viewing the same file.

Adobe has developed a new type of format, called *PDF,* for all types of files. With PDF files, all of the formatting information is stored in the file, and using their specialized viewer, you can view the document as it was created. This new file format is much like Postscript, which is used to print files created in word processors and desktop publishing programs. PDF can also be used for illustrations and virtually every type of file used to convey information. Already, corporations are converting their documentation into PDF to be shared over the Internet, over an internal intranet, and even on CD-ROM.

8.9   SERIAL LINE INTERFACE PROTOCOL AND POINT-TO-POINT PROTOCOL

Serial Line Interface Protocol (SLIP) is used for dial-up connections. It is a protocol that supports TCP/IP over serial communications lines where routers and gateways are not used. If you are using a modem and need to send IP packets over a dial-up line, the transmission will not work because you must connect to another modem, and, of course, modems do not understand IP.

SLIP encapsulates the IP information or information from layers above IP and transmits it over a serial line. Addressing is not used during this part of the connection because a serial line is a point-to-point dial-up communications link. When it is received on the other end, the SLIP data unit is deleted, and the IP data unit can then be sent over a conventional TCP/IP network. The termination point for SLIP connections is usually a gateway computer.

Point-to-Point Protocol (PPP) is a newer version of SLIP that provides faster, more efficient communications. PPP uses a frame format such as High-level Data Link Control (HDLC), with an information field containing the IP header. PPP uses another protocol, the Link Control Protocol (LCP), to establish a connection between either end. Once the connection has been established by LCP, the PPP transmission can begin.

LCP negotiates with the gateway for link configuration, quality of service, network layer configuration, and link termination. The protocol residing in the PPP data unit is identified by LCP at the time of negotiation.

8.10   SIMPLE NETWORK MANAGEMENT PROTOCOL (SNMP)

The Simple Network Management Protocol (SNMP) provides a standard protocol that can be used on any platform. Vendors have already implemented SNMP support in their products, which resolves the issues of interoperability. Modems, routers, gateways, and host computers from different vendors can all share network management data by using SNMP.

SNMP uses simple code, allowing it to be incorporated into devices that do not typically have a lot of processing resources.

Small routers and modems can be equipped with SNMP software without affecting the performance of the device. Yet SNMP is also flexible enough to allow changes to be made easily.

SNMP uses the services of UDP to share data between the various nodes on the network. Network administrators then use client software to access SNMP host software that is responsible for the gathering of network data. The server uses a store-and-forward methodology to maintain information about network entities.

The client application used by the network administrator may provide a simple interface with a graphical user interface (GUI), which is usually the case, or with something simpler and less user friendly. This, of course, is transparent to SNMP, which does not define the user interface.

It is important to understand that SNMP defines the means by which devices communicate network data to one another and not how the data is to be displayed to the user. This is the job of the client application, which collects the data from an SNMP server and displays it in a variety of formats.

The various devices on the network use an agent. The agent reports to managers, which are communications software that query devices for network data. The data includes status of the device, the links attached to the device, and maybe even measurement data used for statistics. This information is then stored in a database on the server. The database is called a management information base (MIB).

SNMP can be used on any type of network and is not limited to TCP/IP. The only requirement is that the network must provide a simple transport service (such as X.25 or ISDN). More complex transports can be used but require additional overhead because of their complexity.

8.10.1   Management Information Base (MIB)

The MIB is stored within the various devices that provide data to the management server. The device stores status and performance data about itself in the MIB, which when queried, provides the data to the manager. The manager then stores the information for later retrieval by the client, which is the user's interface.

The client will then display the information to the user (usually the network administrator) in a variety of formats, depending on the capabilities of the application. The client software used to display the SNMP data is not standardized (nor should it be). The information displayed for each device includes the system identification, the number of interfaces and the status of those interfaces, routing table information (such as available routes and status of those routes), and traffic measurements. Statistics such as the number of cyclic redundancy check (CRC) errors encountered and the number of retransmissions are also common.

The MIB is used to define the data that is to be stored by each entity. The user of the device, such as a network administrator, configures this. Keep in mind the types of devices we are talking about: routers, gateways, modems, and network servers.

9

Signaling System #7

Signaling System #7 (SS7) is a specialized network protocol used for the connection/disconnection of all telephone calls. Without this powerful network and its associated protocols, telephone companies could not deliver the many services they do today, such as 800 services, calling name display, and custom routing features.

When a telephone call is placed, the local telephone switch generates a data message identifying the various aspects of the call and routes this message through the network to an adjacent telephone switch. The intent of this message is to establish a connection between the two switches to route the telephone call. The adjacent telephone switch confirms that the circuit identified is available and reserves this circuit until the connection is actually ready to be made (when the called party answers the phone).

The adjacent telephone switch must then generate a similar data message and send it to the next switch in the connection path. This process repeats itself until an end-to-end connection has been made. The terminating switch then rings the called party's telephone and when it is answered, sends a data message in the backward direction, completing the connection phase. This is an oversimplified explanation of what SS7 is used for; it is much more complex than this.

SS7 is also used for connecting to databases used in telephone networks for routing instructions, user profile data (calling card information, billing information, custom feature definitions, and so on), and specialized applications used by the telephone companies. It may sound like this process of connecting a telephone call would be slow, but in reality, SS7 has had a significant impact on the speed at which telephone calls are placed.

There are some who believe that SS7 is obsolete and is being replaced, especially with converged networks using TCP/IP. However, this could not be further from the truth. SS7 is being modified to support these new technologies, but a new technology to replace SS7 would take far too many years to develop and would end up duplicating the very same functions SS7 provides.

9.1 SS7 ARCHITECTURE

The SS7 network has a fairly simple design, consisting of three main functions. The endpoints originate SS7 traffic, the transfer points are responsible for routing the messages through the network, and the control points are responsible for managing traffic to databases that are resident in the network. Any of these functions can reside as a stand-alone system or as part of a switching function.

9.1.1   The service switching point (SSP)

The service switching point (SSP) is where SS7 messages originate. The SSP function is found most often in the end-office telephone switch. Commonly found as an adjunct processor, the SSP originates SS7 messages after determining which interoffice trunk will be used to connect a call. The SS7 message is sent to the end office on the remote end of the circuit and contains a request for connection.

The SSP also originates an SS7 message when it cannot determine which interoffice circuit to use to connect a call. For example, in the case of an 800 number, the end office cannot determine which circuit to use because it cannot determine how to route the call based on the digits dialed. The 800 number must be converted to a routing number before the switch knows which trunks to use.

The SSP originates a query to an SS7 node that will provide a connection to a database. One of the unique things about SS7 networks and databases is that the SSP does not have to know the address of the database. The message originated must only provide the digits dialed. The signal transfer point (STP), which is responsible for routing SS7 messages through the network, can use this information to determine which database the query should be sent to.

Once the database has provided the necessary routing information, the SSP can begin connecting the necessary facilities to handle the call. Keep in mind that this process must be repeated every time a circuit is needed. It is rare that a central office will have one circuit that can handle calls to remote offices without involving other intermediate switches.

Although it appears that this system would slow down call processing and result in unnecessary delays, the network operates very quickly. There are very few delays in SS7 networks, due partly to the speed of the network and partly to its implementation.

9.1.2   The signal transfer point (STP)

The STP is responsible for routing traffic through the SS7 network. It is not the originator of any traffic (other than network management) and is never the final recipient of any SS7 traffic. It is an intermediate point that provides some processing and routing of SS7 messages.

The simplest traffic for the STP is circuit-related traffic. This traffic is originated by an SSP and sent to another SSP to request a connection on a particular circuit. The SS7 message is addressed to the remote SSP and does not require any processing by the STP. When the message is received at the STP, it is simply passed along without further delay. This is called *ISDN User Part* (ISUP) traffic.

Non-circuit-related traffic usually involves a database. We have already discussed numerous applications that use databases, such as 800 numbers. They almost always require the processing services of the STP.

When a query is made to a database, the SSP does not typically know the address of the specific database it needs to query. This is desirable because of network management considerations. If the databases were to be addressed directly, and the database was unavailable for any reason, there would be unnecessary delay in trying to determine how to best handle the query.

The STP provides global title translation (GTT) to determine where a query should be routed. The STP looks at the Signaling Connection Control Part (SCCP) to determine what digits were dialed (which is part of the called party address) and makes its routing decision based on these global title digits. The STP does not have to look at all of the digits dialed, just the area and office codes.

Global title digits (usually the digits dialed in the called party address) do not have to be a telephone number. In the cellular network, global title digits are the mobile identification number (MIN) used to identify a cellular terminal. GTT digits can be any kind of number.

Global title is a unique feature of SS7, and it is valuable for many reasons. In most networks, the address of an end node must be propagated throughout the network. This would mean advertising the address of all databases (or *subsystems,* as they are called in SS7) throughout the SS7 network. If the subsystem should fail or become congested, the address would have to be marked as unavailable. If the address were propagated throughout the network, the status information would also have to be propagated throughout the network.

With SS7, only the STP knows the database address. Not all STPs know the addresses of all databases. A partial global title can be performed, routing the message to the appropriate STP for final routing. This allows network operators to simplify routing tables in their STPs throughout the network.

Another advantage of using global title is anonymity. In some networks, the subsystem is shared with other networks. The host network does not want nodes outside of its network to be able to address the subsystem directly and will not advertise the actual subsystem address. Instead, all queries are directed to the host's gateway STP, which performs gateway screening first and then GTT to determine the best subsystem for the application.

Another feature of the STP is network security. The STP can be used in a gateway function, allowing only traffic that meets specific criteria to enter into a network. This is an important feature as telephone companies begin to share their network resources. The STP can be configured to allow messages with specific origination addresses or portions of an origination address (such as the network or the node portion of the address).

The STP can also be configured to allow only specific types of SS7 traffic from specific addresses. For example, a screen can be configured to allow a particular network to send ISUP messages but not database queries. This function is referred to as *gateway screening* and is found in "edge" STPs, which connect to STPs in other networks.

The STP also manages the resources of the network. Resources in an SS7 network are primarily subsystems. To connect to a subsystem, the STP must route to a front end, the SCP. The SCP interacts with the STP to ensure that the actual subsystem is available and can process the queries being sent. If a subsystem fails, the STP provides subsystem management procedures that will reroute queries to other subsystems, either redundant or not.

An important function of an STP is route management. The STP receives network management messages identifying problems with other nodes adjacent to the STP. The STP can then reroute messages based on the status of the network. Only STPs provide route management.

The STP function may change as the network evolves over to ATM. STPs will no longer be required for routing messages through the network. The ATM network may have a flatter hierarchy, with all traffic routed by the ATM switch. The STP will still be needed for network management, subsystem management, and gateway screening.

9.1.3 The service control point (SCP)

The service control point (SCP) is a front end to subsystems. The SCP is not a database itself, although it can certainly be collocated with a database. The SCP is a function that manages access to the various databases it attaches to and can manage more than one subsystem.

Each of the databases is addressed through a subsystem number that identifies the application served by the database. The network operator predefines these subsystem numbers, and in some cases, universal subsystem numbers have been predefined to ensure interoperability between networks.

The database itself does not have an SS7 address. Queries must be sent to the address of the SCP. The SCP then routes queries to the appropriate subsystem based on the subsystem number. This offers maximum flexibility, without having to use "locked in" address labels. If a particular subsystem fails, there are no physical addresses to deal with. Subsystem management, which is the responsibility of both the STP and the SCP, manages routing to subsystems during failures.

Two different configurations are used for subsystem routing; solitary and replicated. When a subsystem is classified as solitary, it is the only subsystem to be used for queries. All queries for a specific application must be sent to the one subsystem. In replicated subsystems, two modes are used, dominant and load sharing.

In dominant mode, one subsystem is marked as primary. All queries are sent to the primary subsystem first and to the secondary subsystem only when the primary subsystem cannot handle the query (because of traffic flow or failure).

In load-sharing mode, both of the subsystems share queries. Associated queries (transactions requiring more than one message) are sent to the same subsystem. The protocol used for queries (Transaction Capabilities Application Part, or TCAP) is responsible for managing the segmentation and reassembly of messages.

9.2 SIGNALING DATA LINKS

All of the nodes in the network are interconnected via data links. The most commonly used link in North America is 64 kbps. The links are not usually individual circuits but channels in larger facilities (such as T-3s). Telephone companies can use any existing channel as long as there are at least two channels (for redundancy) to any one node. Another rule that is often overlooked is using channels in completely separate facilities. This means that if a T-3 channel is used between two SS7 nodes, the redundant link must be in a T-3 (or other facility) apart from the first. The cross connects, amplifiers, and channel banks must all be separate for the two links. This level of redundancy guarantees the integrity of the link. If any of the devices used by the data link fail, the link could fail. If both of the redundant links are connected through the same devices, both links fail.

This seems like an obvious situation, and the solution seems just as obvious, yet in many networks this is overlooked. Either the facilities are passed onto different groups who do not understand the need for redundancy in the SS7 network, or the network administrator stops at the SS7 node, leaving the rest of the network design to a different group. The most important concepts to understand in SS7 networks are redundancy, integrity, and diversity.

SS7 data links are labeled according to their location in the network. There is no physical difference between the various labels, but their location may determine how they are used. Network management procedures will differ on some links, all managed by the individual nodes. Almost all nodes require identification of the link by its label so that network management procedures can be invoked properly.

As shown in Fig. 9.1, there are several types of links. Links used to connect an SSP to an STP or an STP to an SCP are called *access links,* or A links. These links provide access into the SS7 network from an SS7 end signaling point.

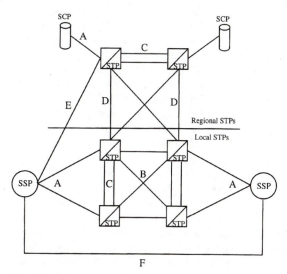

Figure 9.1   SS7 signaling links.

Bridge, or B links, connect STPs within the same network to one another. Some networks use a hierarchical approach to STP deployment, using a pair of regional STPs as hubs. If an STP becomes congested, messages can be routed to the regional STP. These regional STPs are also used for access to network databases. The links connecting to these regional STPs are called diagonal links, or D links, and connect local STPs to regional STPs.

An SSP can also connect to an STP outside of its area. For example, there may be an STP within the same local access and transport area (LATA) as an SSP that cannot be reached. An extended link, or E link, can be used to reach an STP outside of the LATA. When an STP is deployed, it is always deployed in a pair. The two mated STPs interconnect through cross-links, or C links. Normally, traffic does not pass through the C links. Only network management messages are passed over the C links to the mated STP. However, if an STP fails, traffic can be routed from one STP to its mate STP over the C links.

Some SSPs may have a high volume of SS7 traffic to another SSP. In this case, it would be advantageous to provide a direct connection between the two SSPs and use associated signaling rather than quasi-associated signaling. A fully associated link, or F link, provides a signaling connection between two nodes.

Deployment of these links is always in pairs. If a link is provided from an SSP to an STP, another link from the same SSP must be provided to the mated STP. The key to network integrity in SS7 is diversity and redundancy. Diversity means that links are provided on separate facilities (for instance, two different T-3s) rather than on the same facility. This way, if a cable should be cut somewhere in the network, only one link is affected and not both redundant links.

This principle applies to all equipment used to support the link. Channel banks, clocks, power supplies, and repeaters must all be deployed in such a fashion that a single point of failure does not cause an SS7 node to become isolated from the rest of the network. This is the biggest challenge in SS7 network design.

Links terminating to the same signaling point are put into groups, called *linksets*. A linkset is a group of links, all of which terminate to the same node. Links are deployed redundantly, which means if there are two links terminating to an STP, there will be two links terminating to its mate as well. These links can be configured into two separate linksets, or they can be configured as a combined linkset. The configuration is controlled by each signaling point through administrative procedures.

A signaling point uses routing tables to determine which linksets should be used to reach a particular point code. These routing tables define a set of routes. A route is a collection of linksets that can be used to reach a point code. There is no knowledge of intermediate point codes in the routing tables. All the signaling point needs to know is which collection of linksets will get to the designated point code.

There is usually more than one route to get to a point code. Routes can be prioritized so that the signaling point can determine which route should be selected first, which one second, and so forth. A group of routes is called a *routeset*. In summary, a signaling point selects a routeset, which consists of a group of routes, which consist of a group of linksets, which consist of a group of links.

9.2.1   56/64-kbps links

As mentioned before, 56/64-kbps links are the most common configuration used in North America. The interface used is the

DS0A, which is typically derived from a T-1 facility. The telephone office will typically demultiplex the T-1 (or T-3) using a digital cross connect. The individual DS0s are then distributed to different equipment in the central office, including the SS7 nodes. The channels used by SS7 must be dedicated channels that are only used for SS7.

One of the most critical aspects of this interface type is timing. DS0s require synchronized clock sources at both ends of the connection. There are a number of devices used in between the SS7 nodes, such as channel banks, repeaters, and digital cross connects. All of these devices must have a clock connection, which must be synchronized with all of the other devices in the network.

Such synchronization is achieved through the synchronization network. This network uses a highly accurate clock source, which is then distributed to all of the telephone offices in the network. Each office is responsible for distributing the clock signal to all of the devices in its building, using highly accurate clock distribution systems.

Although the 64-kbps link provides more than adequate bandwidth for signaling, the full capacity of the link is protected. The links are configured to carry no more than 40 percent of their capacity. The various SS7 nodes distribute traffic over all of the available links to ensure efficient use of the data links bandwidth.

In the event of link failure, traffic is rerouted over the other available links. This is one of the reasons for limiting the capacity of the data link. When there is a link failure, and traffic is diverted to another link, that link can effectively carry 80 percent of its capacity. Therefore, enough bandwidth is left over to carry network management traffic.

9.2.2 1.544-Mbps links

Although the SS7 network has not yet run out of bandwidth, the type of traffic is changing, and the need for additional bandwidth is changing. At the same time, the telephone network itself is changing, and we are seeing that the present digital hierarchy is quickly becoming obsolete. What this means to SS7 network operators is a change in the type of interfaces they can use.

As the voice network is converted first to fiber optics and then to ATM, the presence of DS0 interfaces will dissipate. SS7 networks will have to convert as well or face additional costs in maintaining separate network facilities. Bellcore issued a recommendation to SS7 network operators, converting networks in two phases.

The first phase involves changing from DS0 links to DS1 links, which provide 1.544 Mbps of bandwidth. This will mean additional capital expenditures to change the existing equipment to this new type of interface. Virtually every circuit card used to connect to an SS7 link will have to be changed. This cost can be rather prohibitive, with no real evident payback.

The second phase replaces all channelized facilities with ATM links. Most telephone companies are deploying ATM links at DS1 speeds to their SCPs first, then migrating their remaining links to ATM. By using DS1 speeds, they can continue using their existing channelized equipment (with modifications to support ATM protocols) until they are prepared to replace their outside plant with ATM switches and fiber optics. The single high-speed link (HSL) is nonchannelized, supporting 1.544-Mbps transmission.

9.3 SS7 PROTOCOLS

SS7 is really a family of protocols, each protocol serving a specific purpose. SS7 uses a layered hierarchy, just as OSI, but because SS7 was developed before the OSI model, there is not a one-for-one correlation. Figure 9.2 shows the protocol stack for SS7.

The packets used in SS7 are called *signal units*. There are three types of signal units used in SS7:

- Fill-in signal unit (FISU)

- Link status signal unit (LSSU)

- Message signal unit (MSU)

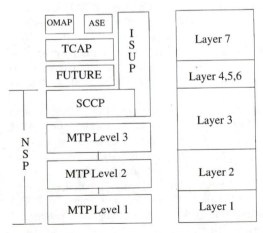

Figure 9.2   SS7 protocol stack.

The FISU is used during idle periods, when there is nothing else to send. The idea of the FISU is to provide enough of a signal unit that link integrity can be checked even when the link is idle and not carrying any real traffic. There is no signaling information in the FISU, only a sequence number and FCS field, as seen in Fig. 9.3. The sequence numbers can be used to send acknowledgments of previously received message signal units; otherwise they carry no significance.

Also notice that there is an absence of addresses in the FISU. This is because they are of local significance only. FISUs are never routed over the network. They are sent from one node to its adjacent node only and are used by the two nodes to verify that the link can still carry traffic. If a particular link cannot carry traffic, the protocol procedures defined for the Message Transfer Part (MTP) are used to reinitialize the link. FISUs are once again sent to verify that the link has been reinitialized properly.

When an excessive number of FISUs have been received in error (determined by checking the FCS field), the link is reinitialized. The signaling unit error rate monitor (SUERM) monitors this activity. This is an incremental counter that counts every signal unit received in error. When 64 signal units have been received in error, the SUERM initiates link alignment procedures.

What is different about this counter is that it also decrements. When 256 signal units have been received without error, the counter decrements by 1. This is sometimes referred to as the

FCS		LI	F I B	FSN	B I B	BSN	FLAG

Figure 9.3 FISU format.

"leaky bucket" technique. Without this feature, whenever the counter reached 64, the link would be realigned, even if unnecessary. The counter is maintained by the link software and is found in all SS7 network nodes. The SUERM is also maintained on a link-by-link basis and is of local significance only. Nothing is shared between two adjacent nodes unless the link fails and is realigned. When this occurs, network management messages are exchanged.

The LSSU is actually part of the MTP protocol. It is generated by level 2 MTP and is used to send link status information to the remote end of a link. The LSSU is not routed through the network; it is only used between two adjacent nodes to manage the links interconnecting them (see Fig. 9.4).

When the link is being realigned, the LSSU is used to identify the various stages of link alignment (such as out of service and in service). This is needed so that the remote end knows when the link is capable of sending traffic again.

Most errors occur because of timing problems and can be rectified by resynchronizing the link with the clock source. This is alignment of the link. For detailed information about how link alignment works and the specific sequence of events, you can refer to my book, *Signaling System #7* (McGraw-Hill),

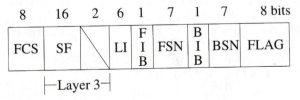

Figure 9.4 LSSU format.

which provides a comprehensive look at the SS7 network and its family of protocols.

The other type of signal unit is the MSU, which is used to carry all other types of SS7 messages. ISUP, TCAP, Telephone User Part (TUP), and network management are examples of the types of messages found in an MSU. Think of the MSU as the signal unit that always carries a payload, whereas the LSSU and FISU carry little or no information. The MSU also has a portion used by level 4. This is the user data portion shown in Fig. 9.5. It is where you will find the application information, such as the upper-layer protocols. The routing label is not part of level 4, but part of level 3, even though it is found in the user data portion of the packet.

The length indicator (LI) identifies the type of SS7 signal unit being sent. There are only three options: FISU, LSSU, or MSU. It does not give the total length of the entire packet, just the length of the field immediately following the length indicator. If the value is 0, the signal unit is an FISU (there is no user data portion in an FISU). A value of 1 or 2 indicates a 1- or 2-byte field, indicating an LSSU. Anything with a value of 3 or larger (up to 63) is an MSU.

SS7 packets can certainly be larger than 63 bytes, but again, it is not the intent of the length indicator to give the length of

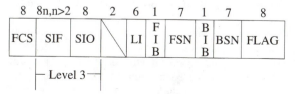

Figure 9.5   MSU format.

the entire packet, only the "user data" portion found after the LI field. It is used by level 2 software in SS7 nodes to identify the type of signal unit being received so it can determine how to handle the signal unit.

9.3.1 Message Transfer Part (MTP)

The MTP is the transport protocol for SS7. The job of MTP is to move SS7 traffic through the network, providing connectionless services. MTP provides sequenced delivery of all SS7 signal units and is divided into three levels. If you look at Fig. 9.2, which shows the SS7 protocol stack, you will notice that MTP includes the first three levels of the stack.

The first level, the physical level, aligns with the OSI model in terms of function. The physical level provides the electrical interface used on the data link and is responsible for the transmission of SS7 traffic over whatever facility is being used.

The data link level also aligns with the OSI model and provides error detection/correction, as well as sequenced delivery of all SS7 signal units. The unusual part here is that MTP provides sequenced delivery when it is a connectionless service. In other data protocols, connectionless services do not support sequenced delivery. In MTP, sequenced delivery is guaranteed because MTP always routes associated signal units over the same path (based on the rotation bits located in the header).

If MTP were a connection-oriented service, a connection between the two ends of the link would have to be negotiated. This would involve messages being transmitted from one node to the other, requesting a connection establishment between the two nodes. This is not the case in MTP, which makes it unique.

MTP level 3 is used for routing and network management; level 2 MTP has no routing information. In fact, look back at the FISU, and you will see the components of the MTP level 2 header. MTP level 3 includes the routing label, which is contained in the user data portion of the header (see Fig. 9.6).

Sequence numbering in SS7 is used to guarantee ordered delivery of all signal units. It is used to acknowledge received signal units as well as to request retransmissions. Sequence numbering uses the backward sequence number (BSN) field, the forward sequence number (FSN) field, and the indicator bits (backward and forward). They are similar to other bit-oriented protocols that use sequence numbering, with a slight twist. The BSN is used to acknowledge the last received sequence number. In other protocols, this field indicates the next expected sequence number.

The FSN is incremented each time a signal unit is transmitted. The only exception to this rule is in the case of the FISU. Unless the FISU is being used to acknowledge received signal units, the sequence numbers remain static, not incrementing. There is no reason to increment the sequence numbers of an FISU since they contain no data. If a signal unit is lost, it will not be retransmitted. FISUs are used for error detection only.

If a signal unit is received in error, the SS7 node level 3 or level 2 (depending on the type of error) rejects it, and the next

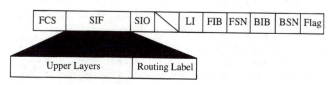

Figure 9.6   MTP format.

signal unit sent in the opposite direction will carry a retransmission request. Retransmission is requested by using the indicator bits. In normal mode, both indicator bits should be the same value (it really makes no difference what that value is).

To indicate a retransmission, the backward indicator bit (BIB) is toggled to an opposite value (being 1 bit long, there are only two options here). When the forward indicator bit (FIB) and the BIB are different values, the node interprets this as a retransmission request. The node will then look at the BSN to determine the last good signal unit received, discard all signal units up to that sequence number from its transmit buffers, and send all remaining unacknowledged signal units from its transmit buffer.

Keep in mind that sequence numbers are of local significance only. In other words, they are not carried and interpreted end to end. As a signal unit moves from node to node, the sequence number changes accordingly and is examined by each node. Several links are involved with any transmission, so this information will not only change every time a message is passed to a new node, but the sequence numbering must be processed at every link. Each link within an SS7 node must maintain its own set of sequence numbers and manage the transmission over that link (this is where distributed processing systems have a real advantage).

9.3.1.1 MTP network management.
MTP level 3 provides network management procedures that reroute traffic around failed or congested nodes. These procedures use the MSU to send messages to adjacent nodes, indicating the status of a node. This is different from the LSSU, which is generated by MTP level 2 to indicate the status of a link between two SS7 nodes.

Three types of network management procedures are used in SS7:

- Signaling link management

- Signaling route management

- Signaling traffic management

Signaling link management controls the activation and deactivation of signaling links. This is controlled by MTP level 3 and will result in an MTP level 2 LSSU being sent to the adjacent signaling point. The alignment procedure is invoked as a result of signaling link management.

Signaling traffic management procedures are used to divert SS7 traffic from a failed link to another link in the same linkset (see Fig. 9.7). They are also used for flow control procedures over a data link. Messages are sent using the MSU signal unit.

Signaling route management procedures are used to identify failed or congested SS7 signaling points. MTP level 3 responds to these messages by rerouting traffic around the indicated point codes. The messages used in these procedures also use the MSU signal unit (see Fig. 9.8). STPs are the only entity within the SS7 network that provide signaling route management. Without an STP in the network, this function cannot be provided.

The procedures used in network management often interact with one another. For example, if a link or series of links fails,

Traffic Management Messages	
0001 0001	Changeover Order
0001 0101	Changeback

Figure 9.7 Traffic management message types.

Route Management Messages	
0100 0001	Traffic Prohibited
0100 0010	Transfer Cluster Prohibited
0100 0011	Transfer Restricted
0100 0100	Transfer Cluster Restricted
0100 0101	Transfer Allowed
0100 0110	Transfer Cluster Allowed

Figure 9.8 Route management message types.

an LSSU (or multiple LSSUs) will be sent over the link to indicate the status to the adjacent node. The LSSU can only be sent if the link will support MTP level 2. If several links fail, the signaling point itself could become isolated or congested, resulting in traffic management procedures and possibly even route management procedures. Each of the procedures works at varying levels, beginning with the signaling link management (which is responsible for individual links).

Next is signaling traffic management, which works with an adjacent signaling point to manage the diversion of traffic from a failed link to a good link. The worst case is route management, where a node is no longer capable of handling SS7 traffic or has become congested to the point where traffic must be diverted away from the entire node. This results in SS7 traffic being rerouted to other signaling points.

As Fig. 9.9 shows, a message is received and examined by the message discrimination function. From here it is determined whether or not the signal unit is addressed to the receiving node or another remote node. If the signal unit is addressed to the receiving node, it is passed to message distribution, which is then responsible for processing the header information and passing the user data on to the next level.

If the signal unit is addressed to a remote node, the signal unit is passed to the message routing function. Here is where the new sequence number and a new BSN (based on the link to be used) are assigned. Remember that the BSN is acknowledging the last signal unit received on that link, not received by that node. This is of some importance to understand since a device such as an STP could have over 500 links terminating to it.

9.3.1.2 MTP addressing. Figure 9.10 shows the routing label and its location. The routing label is found in the user data

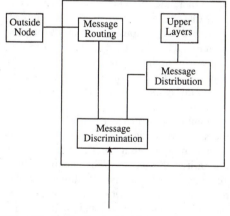

Figure 9.9 Routing functions in MTP.

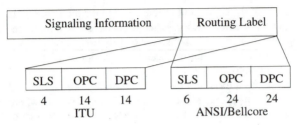

Figure 9.10   MSU routing label.

portion of the signal unit. It contains the actual address of both the source and the destination. The addresses used in SS7 are called *point codes* and consist of different parts, depending on whether the point code is a national or international point code.

The national point code format is different for every country. In the United States there are three parts to the point code, the network identifier, the cluster identifier, and the member identifier. Each of these address portions is three digits long (or 3 bytes long in the binary code). The point codes must be defined by a central body with authority over all point codes in the United States. Currently, Bellcore (now known as Telcordia) is the authorized body responsible for the allocation of point codes to network operators.

The international point code format consists of zone, area, and member identifiers. The zone identifier is a 3-bit field used to identify a country or a group of countries. The area identifier is an 8-bit field that identifies the network. The member identifier is a 3-bit field that identifies the member within a network.

There are two planes, or levels, of networking: the international and the national planes. The international plane uses the ITU standards and allows all countries to interconnect their networks using a common interface. An international gateway STP is used to connect at the international plane, providing protocol conversion from the national protocol (such as the ANSI) to ITU (see Fig. 9.11).

The national plane allows every country to use its own flavor of SS7, without compatibility problems with the rest of the SS7 network. As long as a country can connect to the international plane, it can still interconnect with other networks.

Once access has been gained at the international gateway, any SS7 traffic can be converted by a protocol converter and routed into any international network, regardless of the country. This is important to all SS7 network operators, who must be able to connect to networks anywhere in the world. It is especially important for cellular and future communications networks, where transparent transfer of service is critical.

9.3.2 Signaling Connection Control Part (SCCP)

Database applications used in SS7 do not have point codes and are identified by a more generic subsystem number, meaning that the database does not have a unique address assigned specifically to that logical entity. The physical SCP used to access the application has a point code, but the application residing on that SCP does not. Instead, a subsystem number is assigned that identifies the application type instead of the entity. For example, to identify an 800 database application, the subsystem number 256 would be used. This same subsystem number would be used for all 800 number databases in the network.

To route to a subsystem, the STP determines which database is best suited for the query received (based on the called party address in the SCCP header) and creates an MTP header addressed to the SCP that is acting as the front end to the subsystem. The

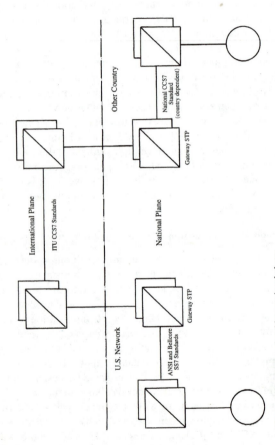

Figure 9.11 National and international planes.

International Plane

ITU CCS7 Standards

National Plane

Other Country

National CCS7
Standard
(country dependent)

Gateway STP

U.S. Network

Gateway STP

ANSI and Bellcore
SS7 Standards

SCP, which may provide front-end services to several subsystems, looks at the subsystem number in the SCCP header and routes the data (minus the SS7 information) to the database.

When the query has been processed by the subsystem, the response is sent to the SCP, which is responsible for placing the data into an SS7 envelope and transmitting the response back to the originating SSP.

The response is addressed directly to the originator of the query, not the STP that provided the GTT. The originating point code is found in the calling party address field of the MSU when the SCP receives it. This portion of the calling party address field is then used to create the header for the response.

SCCP provides several types of services for accessing subsystems. They are divided into connectionless and connection-oriented services. The differences between the services are mostly in the level of service (such as sequencing and segmentation/reassembly). The following are the various classes of service provided by SCCP:

- *Class 0*—Basic connectionless

- *Class 1*—Sequenced connectionless

- *Class 2*—Basic connection-oriented

- *Class 3*—Flow control connection-oriented

- *Class 4*—Error recovery and flow control connection-oriented

Connection-oriented SCCP is not used in the United States but is used in some international networks. In the United States, connectionless services are supported, with some connection-oriented features.

Class 0 services are commonly used in North American networks. They provide delivery of messages, such as database queries, without sequencing. This is similar to other connectionless services found in other protocols. As with all connectionless services, there is no guaranteed delivery.

Class 1 services are also connectionless, with the addition of sequence numbering. MTP sequencing is separate from SCCP sequencing. Remember that MTP is used to deliver SS7 messages from one node to the next. SCCP provides delivery of messages from end to end. Sequence numbers found in SCCP are used when multiple packets are being sent for one transaction. An example of this would be a switch invoking a feature in another switch, where TCAP would send more than one message to the remote switch. Sequenced delivery is required to ensure proper operation.

Even though sequenced delivery is guaranteed with Class 1, it is still considered connectionless because SCCP does not request a logical connection with the remote node. This is somewhat unique to SS7.

Class 2 services provide connection-oriented services, which require a logical connection between the two communicating nodes. A reference number in SCCP is used to identify the logical connection since there are usually multiple transactions to the same node. This is similar to X.25 logical connections.

Class 3 services add flow control and expedited data. Flow control controls the flow of data to a remote node, whereas expedited data identifies messages with a higher priority. This allows SCCP to identify data that must be processed immediately.

Class 4 services (not currently defined in the Bellcore standards but defined in the ANSI standards) provide for error recovery. When an error is detected, SCCP can request a retransmission. This is apart from the MTP error recovery procedure.

SCCP provides end-to-end delivery of signaling data in the SS7 network. Although the standards identify SCCP as providing services to TCAP and ISUP, SCCP can work with any other type of protocol (in theory) and was designed after the X.25 transport protocol.

9.3.3   Transaction Capabilities Application Part

TCAP is used for non-circuit-related signaling, which means that the signaling does not pertain to any one circuit or connection. Examples of non-circuit-related signaling include database queries for 800 services. TCAP is used whenever a database is involved or when invoking a feature in another switch.

TCAP is not limited to wireline services. It is also widely used in cellular networks, which rely on database access for virtually every call placed over the wireless network. These databases contain location information for every active mobile telephone in the network. Before a call can be connected, the location of the mobile subscriber must be determined. This is accomplished through the use of TCAP and cellular protocols (such as IS-41 and MAP).

Cellular switches also communicate using TCAP, providing update information regarding the location of mobile subscribers. The various cell sites in the wireless network send location updates to the mobile switching center in their area, which sends the data to a home location register (HLR) and visitor location register (VLR), which are usually collocated.

TCAP usage was not frequent until telephone companies began providing new services. Its original use was 800 services, but now it has become widespread and includes many different wireline and wireless services. As the telephone network continues to evolve, TCAP usage will become more widespread and will account for the majority of SS7 traffic.

Figure 9.12 shows that TCAP is divided into three different sections. The transaction portion identifies whether or not the following component portion carries a single transaction or multiple ones. It also indicates the type of transaction by providing a package-type identifier, which consists of the following values:

- Query with permission

- Query without permission

- Response

- Conversation with permission

- Conversation without permission

- Abort

- Unidirectional

A query with permission is a database query. Typically, there will be only one query to a database, but there could be multiple associated queries in one transaction. Permission indicates that either end, the originating node (sending the query) or the remote node (the SCP), can terminate the transaction before it has been completed. A query without permission indicates only the origination point can terminate a transaction. This forces the SCP connecting to the database to complete the transaction no matter what happens at the remote end.

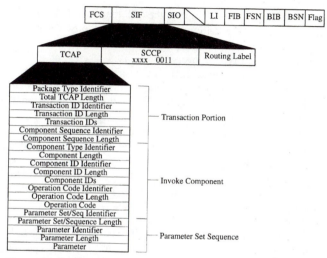

Figure 9.12   TCAP format.

It is probably worth noting here that SCCP subsystem management is capable of detecting problems with subsystems and rerouting queries to backup subsystems in the event of subsystem failure. This is how a query without permission is completed, even when the subsystem being addressed has failed.

The response package type identifies a TCAP message that is carrying the response to a query. This, of course, is what you would see being returned from the subsystem to the originating node. Remember that the MTP destination address for queries is usually the point code of the STP adjacent to the SCP. The STP provides GTT so that the correct subsystem address can be determined. The response will carry the MTP destination address of the originating node, and not the STP that provided the GTT (remember that the calling party address in the SCCP header also contains the origination point code).

The package-type conversation is used when a dialog between two nodes is needed. If a subsystem receives a query, it may be determined that additional information is needed to complete the transaction. The subsystem then has the option of returning a package type of conversation, opening up a two-way dialog between the subsystem and the querying node. This can also be used when two switches need to exchange signaling information while a call is in progress.

The abort package type allows either node (depending on whether or not permission was granted) to terminate a transaction. A reason code is included in the abort message to indicate why the transaction was aborted.

The unidirectional package type is used when a message is sent without a response required. This is the only package type that does not require a transaction identifier. The transaction identifier is used to track transactions sent to a node. This identifier is of local significance only; that is, it is used only by the originating node so that it can correlate responses with queries. When a response (or a conversation) package type is returned, it will contain the same transaction identifier so that the originating node can associate the response with the proper query.

The second portion of the TCAP packet is the invoke component. There may be more than one component in this section. This allows a single transaction to invoke multiple features or send multiple queries (if associated) to the same subsystem.

Each component carries its own unique identifier so that the originating node can keep track of responses to various components. The correlation identifier is found in the component identifier and may be accompanied by an invoke identifier. The correlation identifier is also of local significance only.

The operation code identifies the type of operation to take place. There is one operation code for each component of a transaction. If there are multiple components in a transaction, there will be multiple operation codes. Operation codes are divided into families, with various values for each family. Think of an operation code family as a class of operations. The following are the operation families:

- Parameter

- Charging

- Provide instructions

- Connection control

- Caller interaction

- Send notification

- Network management

- Procedural

- Operation control

- Report event

- Miscellaneous

The operation codes do not provide enough information by themselves and must be followed by the parameters that provide the data needed to complete the operation. Parameters are found in the next section of TCAP, the parameter set sequence. There can of course be multiple parameters for an operation.

As mentioned above, each of the operation codes requires additional parameters. The last section of the TCAP packet consists of those parameters. The organization represented in Fig. 9.12 suggests that all of the fields are organized in linear fashion. This is not the case. The transaction portion identifies how many components are to be found in the component portion. Each component identifies an operation code, followed by its respective parameters in the parameter set sequence section. The component section is then repeated for the next component, followed by its respective parameter set sequence.

9.3.4   ISDN User Part (ISUP)

ISUP is the protocol used throughout North America and many parts of Europe for circuit-related signaling. Circuit-related signaling applies to all voice and data circuits. The ISUP protocol supports both physical circuits and logical channels. This is something its predecessor, TUP, is not capable of supporting. However, ISUP cannot support the virtual circuits and virtual paths used in ATM networks. To support broadband services, a newer version of ISUP, broadband ISUP (BISUP), has been developed.

ISUP provides two types of services, basic and supplementary. Basic services provide support for establishing connections for voice and data circuits in the public switched telephone network (PSTN). Supplementary services are used for the exchange of signaling data that is related to a call already in progress.

There are two methods defined for passing circuit-related information for supplementary services. The signaling information being sent is used by the originating and destination nodes and is ignored by any intermediate nodes along the signaling path. To understand how this works, we should first examine how a call is set up.

As shown in Fig. 9.13, the originating SSP function sends a setup message to an adjacent SSP. This is typically not the terminating SSP and may be a tandem switch. The setup message is routed through an STP. The adjacent SSP then must initiate a setup message to another adjacent SSP (which may or may not be the destination SSP) to establish an end-to-end connection. The use of an STP as the signaling hub is referred to as quasi-associated signaling because the STP performs an intermediate function. When STP is used and the two SSP functions are directly connected, the mode is referred to as associated signaling. This is found when fully associated links (F links) are used (very common throughout Europe today).

It should be noted that the MTP destination address in each case is the point code of the adjacent SSPs and not the final destination. The called party address in the ISUP message is used to determine the destination point code for the next hop. The

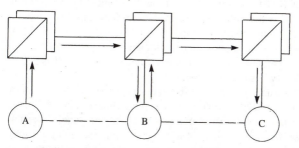

Figure 9.13 POTC call setup using ISUP.

setup procedure must be repeated for each hop used to establish the call connection.

Once a call connection has been established, additional circuit-related information can be sent to the remote SSP. There are two methods used with supplementary services to provide end-to-end signaling once a connection is established and a call is in progress. The pass-along method uses the services of MTP to send ISUP supplementary information, and the SCCP method uses the services of SCCP to send ISUP supplementary signaling. Only pass-along is currently used in North America.

When pass-along is used, the ISUP message is passed along the same path as the call setup messages. This means the message must be sent through all of the intermediate nodes along the connection path. This will, of course, introduce additional delay and is not the best method of transport, but it is commonly used today.

The SCCP method allows the ISUP message to be sent from the originating SSP to the remote SSP, bypassing all of the intermediate SSPs along the signaling path. This is possible because SCCP can route end to end, using STP services to determine the shortest path. MTP services can only support routing from one SSP to another adjacent SSP (using an STP as an intermediate step in quasi-associated signaling). The use of SCCP may become more commonplace as new intelligent network services are implemented.

There are a number of ISUP message types used for the establishment and tear down (release) of voice and data circuits. Figure 9.14 shows how a call from an ISDN telephone (labeled as a terminal) interacts with an ISDN switch (or end-office switch), which in turn generates SS7 messages for the SS7 network.

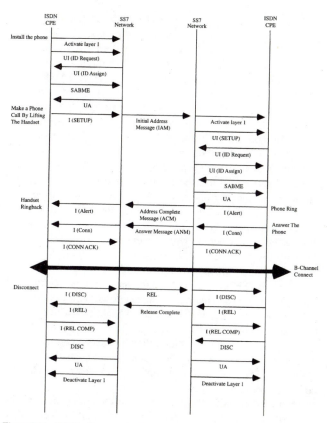

Figure 9.14   ISDN call setup using ISUP.

The initial address message (IAM) is used to request a connection between two SS7 signaling points. The IAM contains both the calling party and the called party telephone numbers. This information is passed through the network without alteration, allowing the final destination SSP to identify the telephone number of the calling party. If this SSP can access a name database, a query can be made to the database, providing the name of the calling party as well. This information is then passed on to the called party telephone display.

If the remote end of the connection is not using an ISDN circuit, but an analog circuit (plain old telephone, POTS, service), a special modem is used at the end-office switch to pass the calling party number and name. A special display at the subscriber's location receives data from the modem between ring cycles and displays the information for the called party prior to the call being answered. This is mentioned so you might understand more about how signaling works. Without SS7, calling name display would not be possible.

When an SSP receives an IAM message, it must determine if it has the resources to maintain a connection (resources meaning processing capacity as well as circuit availability). The circuit to be used is identified in the IAM message as well. However, just because the circuit has been identified by the originating SSP as available does not mean that the same circuit is available at the other end. There are many circumstances that may render the circuit unavailable at one end of the connection.

If the receiving SSP determines that the connection can be made on the designated circuit, it returns an address complete message (ACM). This indicates that the IAM was received correctly and the destination SSP can support the connection. The destination SSP then must determine the circuit to use for the next hop on the connection. Once the circuit has been identified (determined by examining the called party address in the IAM message), the SSP generates an IAM message to the next SSP on the connection path.

Once the circuits have been established end to end, and the called party is alerted (by ringing the telephone), an answer message (ANM) is generated and sent backward from each of the SSPs (back to the originating SSP). The voice circuit up to this point has been only partially connected. Some networks initiate backward cut-through, which means that the voice circuit is connected in the backward direction. This allows service tones (such as ringback and busy) to be sent to the calling party by the remote SSP.

Other networks do not actually cut through the voice circuit until the called party answers. All service tones are provided by the originating SSP. Implementation is network-dependent and varies from one service provider to the next. Almost all Bell Operating Companies (BOCs) follow Bellcore recommendations, which vaguely indicate that backward cut-through should be used.

Once the ANM has been received, the call is considered in progress. The ISDN B channel is now connected (or the analog circuit in the case of POTS service). No other messages will be sent unless supplementary services are initiated.

When either party hangs up, a release message (REL) is sent. It can be sent in either direction. The release message is sent to each of the SSPs, allowing each of the circuits to be released and made available for another connection independently of one another. This is different from other signaling methods, where the connection was maintained end to end until both parties had hung up. By releasing each circuit independently, the circuits can

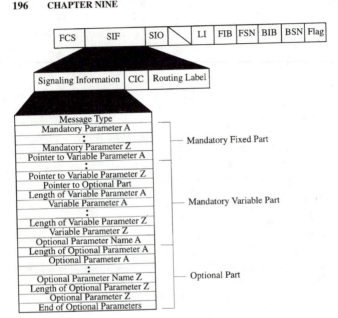

Figure 9.15   ISUP message structure.

be made available for new connections much more quickly.

When an SSP receives the release message (which will also identify the reason for releasing), a release complete (RLC) message is sent as confirmation that the REL was received and the circuit has been released.

The ISUP message consists of three parts, as shown in Fig. 9.15. The circuit identification code (CIC) is also part of the ISUP message and is located after the routing label.

The CIC is not found in BISUP because it cannot support

broadband services. The CIC can only identify narrowband circuits used for interoffice trunking. The identification is a simple number, which is maintained by each node. The circuit identification for any one circuit must have the same identity at both ends of the connection.

The mandatory fixed part contains the message type and a pointer to the next portion of the ISUP message. *Mandatory* implies that this portion of the ISUP message must always exist, and *fixed* indicates that the fields in this section are of a fixed length. The pointer is used to identify where the next section, the mandatory variable part, begins. It provides the offset, which is the number of octets to be counted from the message type to the beginning of the mandatory variable part.

Figure 9.16 identifies the message types used in ISUP and BISUP and whether the message types are used in ITU or Bellcore networks.

The next section of the ISUP message is the mandatory variable part. This section is not usually mandatory, depending on the message type. It is possible to have an ISUP message with only the mandatory fixed part and an optional part. It is also possible to have an ISUP message with only a mandatory fixed part. There can be any number of variables in this section, depending on the message type. It is in this mandatory fixed part that you will find the called party address in the IAM message.

The section following the mandatory variable part is the optional part. This is the most frequently found section in ISUP messages; that is, the majority of ISUP messages use the mandatory fixed part and the optional part. The calling party address is in the optional part.

There are many ISUP parameters that may use additional

parameters, which are found in the optional part. If we look at the overall structure, assuming all sections of ISUP are used, we will find the message type (such as IAM) followed by a parameter or series of parameters. Any one of the parameters could have additional parameters found in the optional part.

ISUP was the first SS7 protocol to be implemented in North American networks. As the SS7 network grew, TCAP services became more prevalent. Future networks will see more TCAP traffic than ISUP traffic as databases become more and more important in the intelligent network.

Address Complete	ACM
Answer	ANM
Blocking	BLO
Blocking Acknowledgment	BLA
#Call Modification Completed	CMC
#Call Modification Reject	CMRJ
#Call Modification Request	CMR
Call Progress	CPG
#Charge Information	CRG
Circuit Group Blocking	CGB
Circuit Group Blocking Ack	CGBA
Circuit Group Reset	GRS
Circuit Group Reset Ack	GRA
Circuit Group Unblocking	CGU
Circuit Group Unblocking Ack	CGUA
Circuit Query	CQM
Circuit Query Response	CQR
Circuit Reservation	CRM
Circuit Reservation Acknowledgment	CRA
Circuit Validation Response	CVR
Circuit Validation Test	CVT
Confusion	CFN
#Connect	CON
*Consistency Check End	CCE
*Consistency Check End Ack	CCEA
*Consistency Check Request	CCR
*Consistency Check Request Ack	CCRA
Continuity	COT
Continuity Check Request	CCR
#Delayed Release	DRS
Exit	EXM
#Facility Accepted	FAA
#Facility Reject	FRJ
#Facility Request	FAR
Forward Transfer	FOT
*IAM Acknowledgment	IAA
*IAM Reject	IAR
Information Request	INR
Initial Address Message	IAM
Loopback Acknowledgment	LPA
*Network Resource Management	NRM
#Overload	OLM
Pass Along Message	PAM
Release	REL
Release Complete	RLC
*Reset	RST
Reset Circuit	RSC
*Reset Acknowledgment	RAM
Resume	RES
*Segmentation	SGM
#Subsequent Address	SAM
Suspend	SUS
Unblocking	UBL
Unblocking Acknowledgment	UBA
*User Part Available	UPA
*User Part Test	UPT
#User-to-user Information	USIS

Figure 9.16 ISUP and BISUP message types.

10

Integrated Services Digital Network (ISDN)

ISDN extends the signaling network already in use by the telephone companies out to the subscriber. This means subscribers can send control information from their PBXs to other PBXs at a remote office, using the ISDN. The ISDN messages are translated at the local telephone company office into SS7 messages and are routed over the SS7 network to the destination exchange, where they are converted back into ISDN messages and forwarded to the destination PBX.

There are many other advantages to having access to signaling information. Consider this scenario. A large corporation sells its products through a mail order catalog. It advertises an 800 number, which routes to its ordering center. When a call is received, the ISDN D channel provides the calling party's telephone number. This is then sent to a database residing in an adjunct processor.

The database looks up the telephone number and finds a record providing the calling party's name, address, and past ordering history (if that person has called in before). The operator then receives the call, along with the database information (displayed on the terminal screen). The operator can now provide more personal service and needs to only verify the address and credit card information because that information is already provided by the database.

The same company can also use ISDN to dynamically route callers based on operator availability. Callers can be routed to customized recordings, voice response systems, and other automated systems, reducing the number of calls requiring a live operator. If the caller has placed an order before, the entire ordering process can be automated by callers entering in the code for the products they wish to order and verifying their address and credit card information via a voice response system.

Signaling aside, there are other reasons for deploying ISDN. Many Regional Bell Operating Companies (RBOCs) are offering Internet access bundled with ISDN service. As the Internet matures and the World Wide Web sites start using interactive software and video, the need for ISDN and other high-speed access lines will be paramount.

Many Internet service providers (ISPs) are offering ISDN service to support the new multimedia sites appearing on the Internet. With new standards from the Internet Engineering Task Force (IETF) supporting multimedia and conferencing (voice and video), as well as programming languages such as Sun's Java, ISDN access to the Internet is becoming unavoidable.

ISDN is more a concept than a technology. The idea is to provide one facility that can support voice, data, and video. ISDN specifications define the protocols to be used to interface the subscriber equipment to the network. Two types of interfaces are supported from the subscriber to the network, basic rate and primary rate. Basic rate interface (BRI) provides two

64-kbps channels and one 16-kbps signaling channel over one circuit. BRI is usually used for residential services or from an ISDN PBX to the desktop. Primary rate interface (PRI) provides twenty-three 64-kbps channels and one 64-kbps signaling channel. The facility is a T-1 facility (DS1). PRI is typically used to connect commercial subscribers with digital PBXs that are capable of supporting a direct ISDN connection.

There are protocols defined for ISDN from the subscriber to the network. The network-to-network protocol is not needed because the voice and other "bearer" traffic can be passed over existing digital facilities. Signaling information is passed from the ISDN signaling channel to the SS7 network, which is separate from the rest of the switched telephone network.

ISDN and SS7 were designed for each other. To be more specific, when the ISDN concept was first conceived, it was determined that a network was needed to support the transfer of signaling information and provide access to intelligent network elements (such as databases). This network was developed before the ISDN interface to the subscriber. ISDN services cannot be provided end to end without full SS7 deployment between all of the participating networks.

10.1   ISDN SERVICES

ISDN supports a number of services from the subscriber equipment to the public telephone network. The concept is to give subscribers one facility and one interface for all of their communications needs. This means ISDN must be able to support not only digital voice transmissions (in the form of telephone calls) but also data transmissions, video (in the form of video conferencing), and packet switching.

Circuit-switching services are those typically related to voice transmissions. The telephone network is a good example of a circuit-switched network. Connections are made by request through the network using switching equipment until a connection has been established from end to end.

Switches are different from data routers because they establish a connection from end to end before actually transmitting anything. Once the circuits have been connected, transmission can begin. When transmission is complete (or in the case of the telephone network, conversation has ended and either party hangs up), the connection is released. Data routers do not establish a connection over circuits but instead route data from multiple sources over the same circuits.

Packet-switched services encapsulate data into envelopes, adding headers containing routing information (as well as other overhead), and transfer the packets through the network using any circuits available. Connections can be established (as is the case with connection-oriented services), but these connections are virtual. A virtual connection means a message is sent through the network to the destination, requesting that resources be allocated for a transmission.

Resources in this sense are processor and software resources rather than circuits. One circuit may be used to transmit packets to a variety of different destinations. Another unique feature of packet-switched services is that transmissions from one source may follow different routes in the network. With circuit-switched networks, all transmissions from one source always follow the same route, using the same circuits.

One of the biggest advantages of ISDN is that both voice and data can use the same facility. Instead of using a dedicated circuit for all data transmissions (which are routed over a separate network from voice transmissions), data can be sent over

ISDN, along with voice transmissions. ISDN is digital and uses time division multiplexing to assign transmissions to channels on a digital circuit. These channels can be assigned voice or data at any time, on demand.

Nonswitched services are typically permanent connections established between two endpoints. These connections are used for the transmission of audio (different from voice, audio is high fidelity, such as that used for radio broadcasts), video, or data. A permanent connection is usually established at the time the ISDN interface is installed at the customer premise.

One of the objectives of ISDN is to support all of these services using a limited number of standardized interfaces. Two interface types have been defined, BRI and PRI. The principal difference between the two interfaces is the amount of bandwidth made available.

One important objective of ISDN is that ISDN services must be transparent. This means that any protocol or application can be used over an ISDN interface, and the ISDN protocol does not affect the transmission in any way. For Internet access, TCP/IP is required to communicate with other nodes in TCP/IP networks. If ISDN is used to access these networks, TCP/IP must be encapsulated into ISDN packets before it can be transmitted over the ISDN interface. The ISDN protocols must not alter or modify any portion of the original TCP/IP packet because this will render it undeliverable in the TCP/IP network. Three services are defined in ITU Recommendations I.200:

- Bearer services

- Teleservices

- Supplementary services

Bearer services include digitized voice, data, and other forms of user data. Bearer services also support transfer of user data using packet-switched services (using X.25). Connection-oriented and connectionless services are supported as well. The ITU Recommendations define bearer service in three parts: the definition of bearer services (I.230), circuit-mode bearer service categories (I.231), and packet-mode bearer service categories (I.232).

Teleservices are defined in the I.200 series as well. Recommendation I.240 defines teleservices, and I.241 describes teleservices supported by an ISDN. Teleservices usually support computer-to-computer types of applications. This includes file transfers, terminal access to remote databases, and other dial-up-type applications.

Teleservices also include teletex (which is ASCII-based text communications via terminal), telefax (facsimile), videotex (enhanced with mailbox functions and graphics), and telex (interactive text communication via a terminal). These are information processing services that are often provided by the telephone company for a fee.

Supplementary services are defined in a number of recommendations. Following is a list of ITU Recommendations that define supplementary services:

- I.250—Definition of Supplementary Services

- I.251—Number Identification Supplementary Services

- I.252—Call Offering Supplementary Services

- I.253—Call Completion Supplementary Services

- I.254—Multiparty Supplementary Services

- I.255—Community of Interest Supplementary Services

- I.256—Charging Supplementary Services

Supplementary services are best defined as enhancements to bearer services. Users of PBX equipment probably recognize many of these services as inherent features of their PBX. Telephone companies are unable to offer many of these services using existing equipment because the equipment is not capable or they cannot obtain the proper tariffs to offer these services. With ISDN, they can offer these services and much more.

Number identification includes calling party identification, both by telephone number and name (calling name delivery). Call offering includes services such as hunting, forwarding, and call transfer. Call completion includes services such as call hold, call waiting, and call busy. Multiparty services are teleconferencing and three-party services. Community of interest is also referred to as a closed user group, which is analogous to extensions from a PBX. Charging includes credit card calls, reverse charging, and usage-based charging.

ISDN protocols are message-based. This means that each packet of information carries a specific meaning, or message, that is to be acted upon by the recipient. The messages used in ISDN protocols provide the instructions and parameters necessary to deliver services to the subscriber.

10.1.1 End-to-end signaling with Digital Subscriber Signaling System No. 1

End-to-end signaling allows companies with private networks to send signaling information from their switching equipment to switches in other remote locations. The signaling information is sent from the ISDN interface (using the D channel) through the public network's signaling network, SS7. Not all telephone companies provide this service, but this was one of the original reasons for providing ISDN. Digital Subscriber Signaling System No. 1 (DSS1) is commonly used by carriers offering services such as prepaid calling because the switches they use to deliver these types of services use ISDN interfaces.

End-to-end signaling allows a PBX to send the same information used in routing, call offering, call completion, multiparty, community of interest, charging, and other call information to their other switches transparently through the public network. This is of value to large corporations with multiple locations.

DSS1 is a collection of protocols that are used to support end-to-end signaling. Q.930, Q.931, and Q.932 make up this signaling system. DSS1 is the signaling used from the user to the network, where it is transferred to the SS7 network for transport to the remote ISDN. Once received at the remote end, DSS1 is again used to deliver the signaling information to the remote PBX.

10.1.2   Private intelligent networks

Private intelligent networks can be of benefit to large corporations because they allow them to more efficiently handle calls to their networks. Before ISDN, corporations used tie lines to connect to their remote switches. The switches often had a star configuration, with one PBX acting as the hub.

The hub was used to route internal calls to other locations, without the users having to dial through outside lines. This let office personnel call other offices by simply dialing an extension number, even when the extension was in another PBX. Complex routing tables and numbering plans had to be programmed into the switches to allow them to provide this capability. Outgoing calls could be routed through the hub, which would then use routing tables to determine which PBX in the private network should be used for the outgoing call.

Corporations could realize big savings because what would normally be a long distance call could be routed to another switch over tie lines, where it would be treated as a local call. Unfortunately, the only information that could be provided to the remote switch was the dialed digits, which were transmitted over the tie lines using tones (such as dual-tone multifrequency, or DTMF).

With ISDN, tie lines are no longer necessary. All of the switches are interconnected to each other using the public network. The private network becomes a virtual private network because there are no dedicated circuits interconnecting the switches. With end-to-end signaling, information regarding call handling can be sent over the signaling channel of the ISDN interface and then transported through the SS7 network to the remote switches.

With digital protocols, there are many more possibilities in messaging. Sending information using tones is very limited because of the possible combinations. In digital information, an unlimited amount of information can be sent regarding virtually anything. Think of reservation centers that receive millions of calls per day. They may have several locations nationwide receiving those calls (such as hotel and rental car agencies). When one particular reservation center becomes congested from too many calls, signaling information can be sent to the main switch to indicate that calls should be routed to another location until congestion subsides. This is achieved using end-to-end signaling.

Telephone companies are now offering intelligent services to their larger customers, in place of ISDN solutions. For example, telephone companies use complex databases that reside in their own signaling network to store routing information for a company. Callers may dial an 800 number, and the telephone company makes decisions about where the call is routed. The routing is based on the call volume for each of the company's reservation centers, or it can be based on the zip code of the calling party.

10.2   ISDN NETWORK ARCHITECTURE

Two interfaces are defined for user access to the ISDN: BRI and PRI.

BRI is a digital circuit, providing two "bearer" channels (both at 64 kbps) and one signaling channel (at 16 kbps). A bearer channel is one in which user data can be transmitted. Only two wires are needed from the central office to the subscriber equipment to support these three channels.

The BRI is a bidirectional interface. In a two-wire circuit, an encoding scheme called *two binary one quaternary* (2B1Q) is used to transmit over one wire and receive over the other. Most BRIs today use 2B1Q because it allows ISDN to be delivered over existing twisted pair. The connection used at the interface is an RJ-45, where pins 3 and 6 are used for transmission and pins 5 and 4 are used to receive.

The BRI supports point-to-multipoint configurations, which is consistent with the intent of the BRI: to support small businesses and residential services. Up to eight devices can be supported by the terminal equipment (TE) on a passive bus up to 200 m from the network termination 1 (NT1).

PRI provides 23 bearer channels (or B channels) and one signaling channel (the D channel), all at 64 kbps, for a total of 1.544 Mbps in U.S. networks and 2.048 Mbps in European networks. The signaling channel supports signaling for all 23 channels.

PRIs are used in commercial applications, usually where a digital PBX exists. If the digital PBX supports direct ISDN connections, the PRI can be terminated to the PBX line card. If the PBX does not support ISDN, a channel bank is required. The channel bank is responsible for receiving the ISDN messages and converting them to analog circuits to be connected to the PBX. This means that the channel bank must be an ISDN-compatible device.

PRI only supports point-to-point configurations, which has been its objective. The original intention of this interface was to support switching devices such as PBXs on the ISDN.

The physical connection to ISDN PRI is a 4-, 6-, or 8-pin modular connector, similar to those found in your home. In the United States, the connector is an 8-pin RJ-45, providing one pair for transmit, one pair for receive, and two pair for power. ISDN telephones receive power either from the network or from an external source. For PRI applications, the power is provided from the telephone company network. In BRI applications, power is usually provided by an external source (such as a power adapter). The phones operate on −48 volts direct current (VOC).

10.3   CHANNEL USAGE

The B channel (also referred to as the bearer channel) is used to transmit digital voice (using 64 kbps PCM). The B channel is also used for high-speed data (either circuit-or-packet switched), facsimile, and slow-scan video. There is not quite enough bandwidth to support broadcast (full motion) video.

The D channel (signaling channel) is used for signaling (basic and enhanced) for all of the B channels on the ISDN interface. The D channel can also be used for low-speed data (such as videotex or terminal communications) and telemetry. Telemetry is used for emergency services (such as alarm monitoring) and allows utility companies to read electrical and water meters. This application has not been widely accepted, especially when cellular services can provide the same capability cheaper than ISDN.

The BRI channel supports multiple devices over the 2 B channels, which means there must be arbitration to determine which device can use the D channel first. The layer 1 protocol handles this arbitration. The BRI is not activated permanently, so there is also a need for activation/deactivation bits. The BRI is only active when a device wishes to transmit.

The PRI D channel supports multiple channels (24 in North America and 32 in Europe). In North America, the D channel is always channel 24. In Europe, the D channel is always time slot 15. Time slot 0 is used in European interfaces for framing. The interface is activated permanently, so there is no need for activation/deactivation bits. The PRI only supports point-to-point configurations, so there is no need for arbitration on the D channel.

Individual channels can be combined to provide additional bandwidth for applications such as video and high-speed access. H channels are available in multiple configurations. An H0 channel supports data at up to 384 kbps. Two H1 channels, H11 and H12, provide 1.536 and 1.92 Mbps, respectively. These are used for high-speed data, near-broadcast-quality video, high-fidelity audio, and voice applications.

Some applications do not require the 64 kbps provided in a B channel. Terminal applications may only require a 9600-baud connection. When this is the case, rate adaption is used. With rate adaption, only the portion of the channel needed is used, and the rest of the channel is filled with binary 1s. Transmissions can be multiplexed (interleaved) with other data over the same channel, allowing the full bandwidth to be used.

10.4   THE NODES AND THE REFERENCE POINTS

ISDN standards identify the functions to be provided at various points in the ISDN. The point between each of these functions is a reference point. There is really no physical entity associated with reference points or ISDN functions. A device can provide one function or several functions and can bridge more than one reference point. Figure 10.1 shows the ISDN Reference Model. Each of the functions is identified in the boxes, and the lines between the boxes represent reference points.

NT1 isolates the user equipment from the local loop. This

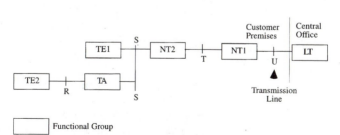

Figure 10.1   ISDN reference model.

allows the telephone company to use digital loop techniques (such as 2B1Q encoding) without extending their encoding methods into the customer premise. Customer premises equipment (CPE) connects to a standard connector, usually at the NT1. The NT1 provides the functions of the physical layer, such as 2B1Q. From the NT1 to the NT2, a four-wire circuit is used (2B1Q is a two-wire circuit).

Network Termination 2 (NT2) can support up to layer 3 if required. Switching functions and line concentration can be supported if needed. Switching functions are used when connecting ISDN to a PBX. Line concentration is used when connecting multiple ISDN telephones or terminals to one ISDN connection (which is the case with BRI and residential services).

Terminal Equipment 1 (TE1) represents functions provided by ISDN telephones and terminals. These are digital telephones, modems, and other devices that can support the ISDN protocol and connect to the ISDN interface at the NT1. Non-ISDN devices must use a different interface that provides ISDN support as well as analog-to-digital conversion.

Terminal Equipment 2 (TE2) is equipment that is not ISDN compatible. The functions provided here support analog telephones, modems, and terminals using RS-232 serial connections. Devices used to connect to X.25 networks are supported by TE2 as well.

The terminal adapter (TA) supports the connection of TE2 equipment to ISDN circuits. The TA is usually a function provided by devices supporting NT2 functions, such as an ISDN telephone. Many of these devices have a connection for interfacing with non-ISDN equipment.

10.4.1   ISDN reference points

The user (U) reference point separates the network side of the ISDN from the subscriber. This is a full-duplex subscriber line. The U interface is not well defined in ITU Recommendations but is defined in the ANSI standard T1.601. This standard defines the U interface using 2B1Q encoding, supporting full-duplex operation on one pair of twisted pair cable. 2B1Q uses four voltage levels, which permits 2 bits of information to be conveyed by each "pulse." Each signal has four possible values, which is how 2 bits can be represented per pulse.

The terminal (T) reference point separates the network termination equipment from the user's termination equipment. For example, with BRI the subscriber loop is a two-wire circuit with 2B1Q encoding. The NT1 on the user side supports a four-wire circuit, with separate transmit and receive pairs.

The system (S) reference point separates the user terminal equipment from network functions. The user TE does not need to be concerned with network functions; it is only concerned with applications. This reference point represents that separation. The rate (R) reference point separates non-ISDN equipment from adapter equipment. Again, the purpose of establishing a separate interface for connecting to ISDN (instead of connecting CPE directly to the ISDN circuit) was to provide a means by which the network could continue to evolve without affecting the CPE. The NT1 isolates the CPE from changes made in the subscriber loop.

10.5   PROTOCOLS OF ISDN

ISDN uses several protocols to support various functions within the network. In this section, we will examine these protocols and the functions that they provide. As with the rest of this book, this section will provide an overview of these protocols rather than a detailed description of each.

Protocols used in ISDN differ according to the layer of services and the type of bearer services. ISDN layers do not go beyond the network layer since this is a transport technology for user information. The protocols and the ITU Recommendations that define them are listed below:

- Physical Layer—I.430 (BRI) and I.431 (PRI)

- Data Link Layer—I.441/Q.921 (Lap Access Procedures on the D channels, or LAPD)

  I.465/V.120 for circuit switched and semipermanent on B channel
  Lap Access Procedures Balanced, or LAPB (X.25) for packet switching on B channel

- Network Layer—I.451/Q.931 Call control (signaling) on D channel

  X.25 packet level for packet switching on D channel
  X.25 packet level for packet switching on B channel

10.5.1 Link access procedure for the D channel (LAPD)

Layer 2 is supported by the LAPD protocol. LAPD is an HDLC protocol similar to LAPB, which is used in X.25. In fact, LAPD was derived from the LAPB protocol used in X.25 packet-switching networks. LAPD provides error detection/correction, flow control, and addressing. The addressing in LAPD is different from the addressing used in X.25, for obvious reasons. ISDN has different address requirements than X.25 packet switching.

LAPD provides unacknowledged information transfer and acknowledged services. Unacknowledged is the same as datagram services, or connectionless. Acknowledged is the same as connection-oriented. Both acknowledged and unacknowledged can be supported simultaneously over the same D channel because they are sent in correspondence to specific B channels (one B channel can use unacknowledged while others are sending acknowledged).

LAPD addressing supports multiple data link connections. This is different from LAPB, which only supports single data link connections. This is important to ISDN because LAPD must be able to address many different connections at the same time and maintain those connections.

The LAPD frame is shown in Fig. 10.2. The flag is a keep-alive signal, used to maintain synchronization on the link. Like other HDLC protocols, the flag is a pattern of a 0, six consecutive 1s, and a 0 (01111110). When the link is idle, LAPD transmits nothing but flags to maintain timing on the link.

To prevent data from duplicating this pattern, bit stuffing is used. The transmitting node (at layer 2) generates the flag and then begins transmitting a frame. If it detects a pattern of five consecutive 1s in the bit stream, layer 2 inserts a 0 after the fifth binary 1. The receiving node detects the flag and then reverses the process. Whenever a pattern of five consecutive 1s is detected, layer 2 at the receiving node discards the 0 bit that follows. If the receiver detects that the sixth bit is a 1 and the

8	16	1 - N	16	16	8
Flag	CRC	Info	Control	Address	Flag

► First bit

Figure 10.2 LAPD frame format.

seventh bit is a 0, it is accepted as a flag. If both the sixth and the seventh bits are a 1, the frame is considered an abort frame.

Two levels of addressing are provided to identify the end node as well as the service. For example, several terminals may be sharing the same ISDN line, and each terminal may be sending a combination of packet-switched traffic and signaling (both on the D channel). The address used in LAPD identifies both the terminal endpoint and the service to be accessed. This is provided in the LAPD address field.

The address field is a combination of two elements: the service access point identifier (SAPI) and the terminal equipment identifier (TEI). The combination of these two elements is called the data link control identifier (DLCI). This is the same as in Frame Relay.

The SAPI identifies a layer 3 user of LAPD. For example, if a frame is carrying X.25 data, the SAPI value is 16. This SAPI value indicates to the receiver that the payload is X.25 data and will require the services of the X.25 resources available at the receiving node. The SAPI values presently defined are

- SAPI 0—Signaling and call control procedures

- SAPI 1—Packet-mode switching using Q.931 procedures

- SAPI 16—X.25 packet-mode services

- SAPI 32-62—Frame Relay services

- SAPI 63—Layer 2 management procedures

- All others—Future standardization

SAPI 1 is recent, supporting Q.931 signaling with packet-mode services for applications such as user-to-user signaling (end-to-end signaling). SAPIs are unique within a terminal (the same SAPI cannot be used twice with the same TEI on different connections).

The TEI identifies a connection endpoint within a SAPI. The following TEIs are defined in Q.921:

- 0-63—Nonautomatic TEI assignment

- 64-126—Automatic TEI assignment

- 1-27—Group TEI

Nonautomatic TEI assignment is used when the user configures the TEI address into the equipment at implementation. This means when the equipment is initially connected to the network, the TEI must be assigned through administration procedures in the equipment (which may be through a terminal or switch settings). A manufacturer can also assign the TEI to the equipment when it is manufactured.

Automatic TEI assignment allows the network to assign an available TEI at the time a connection is requested. Each time the connection is released, the TEI is assigned to another device that is requesting a connection. A device can have more than one TEI, depending on what type of device it is. A terminal concentrator has multiple TEIs. Figure 10.3 shows the LAPD address field.

The control field carries a control command. These are similar to X.25 control commands. As seen in Fig. 10.4, the control commands are divided into three groups: information, supervisory, and unnumbered. Each group uses its own format.

Information control frames are used for sending user data. They may also carry an acknowledgment of previously received information frames. Information frames are only sent

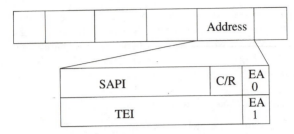

Figure 10.3 LAPD address field.

1 2 3 4 5 6 7 8 9 10 11 12 13 14 15 16

| 0 | N(s) | P/F | N(r) | Information Transfer |

| 1 0 | S S | 0 0 0 0 | P/F | N(r) | Supervisory |

| 1 1 | MM | P/F | MMM | Unnumbered |

N(s) - Now sending
N(r) - Need to receive
S - Supervisory function bit
M - Modifier function bit
P/F - Poll/Final bit

Figure 10.4 LAPD control field.

when there is user data to be transferred across the ISDN.

Supervisory frames are used for flow control and error detection. There are three possible commands used in supervisory frames, receiver ready (RR), receiver not ready (RNR), and reject (REJ). When there is no user data to send, a supervisory frame is used to acknowledge received information frames.

Supervisory frames are also sent to maintain a connection.

RR is used to acknowledge received frames, RNR sends an acknowledgment but also requests the suspension of transmission for flow control purposes, and REJ indicates that the last frame received was rejected and needs to be retransmitted. The N(s) field provides an acknowledgment of the last good frame received. All sequences sent after the sequence indicated in the N(s) field are retransmitted.

Unnumbered frames are used for connectionless services. These frames do not use sequence numbers (which is why they are referred to as unnumbered). There are several commands associated with unnumbered frames:

- Set Asynchronous Balanced Mode Extended (SABME)

- Disconnect Mode (DM)

- Disconnect (DISC)

- Unnumbered Information (UI)

- Unnumbered Acknowledgment (UA)

- Frame Reject (FRMR)

- Exchange Identification (XID)

The SABME command is used to establish a connection with an endpoint. This command is only used with acknowledged (connection-oriented) services. Connection establishment begins with a request for a connection. The SABME sets sequence numbering to 7 bits (0–127) and establishes a logical connection between two peer entities. This is different than in most data communications, where there is no peer relationship but a master/slave relationship (data terminal equipment/data circuit terminating equipment, or DTE/DCE).

One important note about sequence numbering is that it begins at 0 for a connection and is incremented sequentially with every frame sent. This continues for the life of the connection. The numbers are significant to the connection only. When a connection is established, sequence numbering is reset to 0.

The DM command is used to reject a connection request. A connection can be denied for any number of reasons. The DM command identifies the cause for denying the connection (although the cause codes are not very specific).

The DISC command is used to indicate a released connection. Either the originator of the connection or the other endpoint can send it. The DISC will release all resources dedicated to the connection.

UI commands are used by TEI management. When a TEI number is requested by an endpoint, a UI with the assigned TEI value is sent in response. TEI management will be discussed more fully later.

A UA is sent as an acknowledgment without any sequence numbering. The UA is sent in response to a SABME. The UA indicates that a connection has been established and transmission of information frames can begin.

A Frame Reject (FRMR) is sent when a frame is received that violates protocol. There are several reasons for an FRMR:

- Undefined or nonimplemented control field (according to the standards)

- Incorrect length for supervisory or unnumbered frames

- Invalid N(r) (must be between last acknowledged and last sent)

- Information field in I frame exceeds maximum established length

The FRMR will also return the control field of the frame in question as a reference for the receiver.

The XID command is used to exchange DLCI addresses with another endpoint. The XID can be sent as a request or a response. In other words, if an endpoint requests the ID of another endpoint, it sends an XID. The endpoint receiving the XID will return its ID in another XID.

The last field in the LAPD frame is the FCS field. This is similar to other error-detection schemes, where an algorithm is used on the contents of the header or the entire frame and the results placed in the FCS field. When received, the same algorithm is used, and the results are compared to the FCS field received. If there is a match, the frame is considered good. If there is no match, the frame is discarded.

10.5.1.1 LAPD sequencing. Sequence numbers are used for connection-oriented services. The sequence numbers are started when a connection is established. They always begin at sequence number 0. Looking at the format of the Q.921 control field, the sequence numbers can be found in two forms: N(s) and N(r). The N(s) can be considered as "now sending." In other words, the sequence number is assigned to the frame in which it resides.

The N(r) can be considered as "need to receive." This is an acknowledgment of previously received frames. For example, if you send me frames with sequence numbers 1, 2, 3, 4, and 5, I will acknowledge with a N(r) value of 6 (assuming of course I received all five frames).

Since ISDN is asynchronous in nature, it is not uncommon to receive acknowledgments for some frames and not for others sent afterward. There is usually a delay in the acknowledgments, which is why timers are used in all network nodes. The timers are set when the frame is transmitted, and if the timer expires before an acknowledgment is received, the frame is retransmitted. Remember that all transmitted frames are kept in the transmit buffer until they are acknowledged.

As mentioned before, sequence numbers are always reset to zero when a connection is established. This ensures that both endpoints of the connection know where they are starting. LAPD differs from LAPB because it uses a modulo of 128. This means that sequence numbers run from 0 to 127 (compared to LAPB, which uses modulo 8, or sequences 0 to 7).

10.5.1.2 LAPD management. There are two types of management at layer 2: TEI management and parameter negotiation. When a user obtains an automatic TEI assignment, a UI frame with an SAPI of 63 and a TEI of 127 is sent. The information field contains a message type of identity request and a reference number.

In response, the network returns a UI with a message type of the identity assigned. The reference number is the same as the one sent in the request. The assigned TEI value is returned in the information portion of the frame.

Some LAPD parameters can be negotiated between entities. An XID frame is used to change these parameters. The parameters set the timers and counters used for connection establishment, flow control, and error detection/correction.

10.5.1.3 LAPD connection establishment. Before layer 3 can establish a connection end to end, a data link connection must first be established between two adjacent nodes. This must be repeated at every connection point along the path. LAPD uses the SABME message to request a layer 2 connection to an adjacent node, which replies with a UA frame. Once the UA has been received, LAPD can begin transmitting information frames, which will carry the layer 3 messages and user data (once connection establishment has been achieved at layer 3). So there are really two connection establishment procedures: one for the data link layer between each connection point and one at layer 3 providing the end-to-end connection through the network.

10.5.1.4   LAPD flow control and error detection/correction.   When an entity becomes busy, it will send an RNR to an adjacent node. The entity receiving the RNR will then periodically send an RR with the poll bit set to 1. The busy entity must then respond by sending either an RR or RNR (still busy).

When a node sends an RNR, it means that the link is busy, not the entire node. Remember we are still at layer 2 here, and layer 2 only deals with data link procedures. If the entire node were congested, layer 3 management would be responsible for advising all nodes in the network of the busy status.

When a link becomes busy, it means that either the buffer is full or the link cannot process the data fast enough. Think of a printer connected through a serial interface to a computer. The leads on the serial interface are used to either stop the transmission of data to the printer or start the transmission again. This is flow control. The same is true for the data link layer in ISDN. The LAPD protocol provides flow control procedures using the supervisory frame and its associated commands at the link level only.

Timers are used for error detection/correction. During the connection phase, timers and counters are reset. For example, the T-200 timer is started by the initiating entity (the one requesting a connection) after layer 2 transmits the SABME. If the timer expires before a UA is received, another SABME is transmitted. This is repeated until a UA is received.

A counter is also used to prevent layer 2 from constantly sending SABMEs with no success. The counter keeps track of the number of attempts made by layer 2 to establish a data link connection, and after n number of tries, the connection phase is aborted. Network management is made aware of the unsuccessful attempts, and network management procedures are then used to determine and report the fault.

Another use for timers is error recovery. If an entity detects an error in an information frame, it discards the frame. It cannot send an REJ because it does not know where the information frame came from (the DLCI cannot be determined). There is no possible way to alert the sender that the frame sent was found in error.

The sender of the information frame uses a timer to determine when information frames should be retransmitted. Timer T200 is used for retransmission. When an information frame is transmitted, T200 is set to zero. If an acknowledgment has not been received before T200 times out, it sends an RR to determine the status of the adjacent node.

The receiver of the bad information frame will send an RR or an RNR, depending on the state of the link, with an acknowledgment of the last good frames received. The originator then discards all of the acknowledged frames from its transmit buffer and retransmits all unacknowledged frames.

When sending unnumbered frames, LAPD does not provide any error correction or flow control. Error detection is provided, but frames found in error are simply discarded without retransmission requests. LAPD relies on the upper layers to manage retransmission of unnumbered frames.

10.5.2   B Channel Data Link Protocol

ITU Recommendation I.465 defines a data link protocol used on the B channel. The LAPD protocol is only used over the D channel. The protocol for the B channel is also known as V.120. The V.120 protocol allows TE2s to communicate with TE1s, or two TE1s with one another. This means the protocol is used at reference points R, S, and T.

The V.120 frame was derived from the LAPD protocol. When you look at the frame in Fig. 10.5, you will see that they are almost identical. The address field is different from the LAPD address field. Instead of using the TEI and SAPI, V.120 uses a logical link identifier (LLI). The address format is shown in Fig. 10.6.

There is also an information field, which is used to carry the user data (see Fig. 10.7). The information field also has a header, which contains control information necessary to process the contained data. The first part of this header is the TA header. This is an 8-bit optional field that is used when terminal communications is being used. Terminal communications supports applications such as telex, where ASCII characters are sent to a dumb terminal. There is also an optional header extension for control state information. The control state (CS) header is used for flow control. The values are shown in Fig. 10.8.

The address field contains the LLI, which is the data link address used over the B channel. The values for LLI are shown below:

8	16	8or16	8or16	16	8bits
Flag	Address	Control	Information	FCS	Flag

Figure 10.5 V.120 frame format.

7	6	5	4	3	2	1	0
Logical Link Identifier (LLI0)						C/R	EA0
Logical Link Identifier (LLI 1)							EA1

Figure 10.6 V.120 address field.

8bits	8	Variable
H	CS	V.120 Information

Figure 10.7 V.120 information field.

1	1	1	1	1	1	1	1bit
E	BR	Res	Res	C2	C1	B	F

Figure 10.8 V.120 terminal adaptation header/CS header.

- 0—In-channel signaling
- 1 to 255—Reserved
- 256—Default LLI
- 257 to 2047—For LLI assignment
- 2048 to 2190—Reserved
- 8191—In-channel layer management

As seen in Figs. 10.7 and 10.8, only LLIs 257 to 2047 are available for actual assignment. LLI 256 is used for connectionless services. For connection-oriented service, a logical connection must first be established. This can be done over the B channel, but if SS7 is available end to end (and user-to-user signaling services are supported), the D channel is commonly used.

The bits in the control state header are used to communicate lead transitions on modems. For example, most modems are connected using an RS-232C or V.35 interface. The bits in the control state header communicate the lead transitions from these interfaces over the ISDN interface. Not all of the interface leads are represented.

10.5.2.1 V.120 connection establishment.

If connection establishment messages are to be sent over the B channel, the messages are encapsulated in the V.120 frame and sent with an LLI of 0. If they are sent over the D channel, the messages are encapsulated in an LAPD frame and sent with an SAPI of 1. The messages sent over the B channel are similar to those used at layer 3 over the D channel: CONNect, SETUP, RELease, and RELease COMPlete. These messages are not part of V.120 since V.120 is the layer 2 protocol.

The calling or the called party can make LLI assignment. If the calling party is assigning the LLI, an available LLI within the range of 257 to 2047 is included in the SETUP message. If the called party is assigning the LLI, the address is provided in the CONNect message, which is the response to a SETUP message. Release procedures are the same as LAPD, using the RELease and REL COMPlete messages at layer 3.

The V.120 protocol allows multiple logical connections to be established on one circuit between two end users (but not multiple users). The protocol is used for non-ISDN terminal devices (TE2s) connecting to a TA, which in turn connects to a TE1 or NT1. The protocol is not used over the public network, only the user network and only between the devices described above.

10.5.3   ISDN layer 3

So far we have only covered layer 2 protocols and procedures for ISDN. Layer 3 is responsible for the end-to-end connection establishment and transfer of user data. There are also management procedures provided by layer 3.

ITU Recommendation I.451/Q.931 defines the layer 3 protocol used for signaling and control. The protocols defined in this and other related recommendations make up the DSS1 specification. The Q.931 protocol was developed for use over the D channel to support both functional terminals (intelligent devices) and stimulus terminals (digital telephones). Functional terminals are capable of using the full range of Q.931 message types and their parameters, whereas stimulus terminals only use a subset of these parameters.

The message structure of Q.931 can be seen in Fig. 10.9. It consists of a protocol discriminator, a call reference value, and the message type. The protocol discriminator identifies the frame as containing a Q.931 formatted message (versus a X.25 or V.120 message). The call reference value identifies the B channel the message is associated with. If this is a BRI interface, only 1 octet is needed (because there are only two B channels). If this is used over a PRI, 2 octets are needed (to support 24 or 32 channels).

The message type identifies what control message is being sent. A complete list of message types is provided in Table 10.1. The remainder of the frame depends on the message type, since each message type provides particular parameters.

| 1 | 2 | 3 | 4 | 5 | 6 | 7 | 8 |

Protocol Discriminator							
Length of call reference value			0 0 0 0				
Call reference value							Flag
Call reference value							
Message type							0
Other information elements							

Figure 10.9   Q.931 message format.

ISDN addressing consists of an ISDN number and an ISDN address. The ISDN number identifies the subscriber's connection to the network (at the T reference point). The ISDN address identifies an ISDN terminal (at the S reference point). This is layer 3 addressing, which is used end to end (unlike LAPD addressing, which is used link to link). There are many methods for addressing, depending on the network configuration. The ISDN address structure is shown in Fig. 10.10.

The country code identifies the country in which the message originated or a geographical area. A variable number of digits (1 to 3) defined in ITU Recommendation I.163 are used. The national destination code is also a variable-length field. It can be used to reach a network within the country code or to route to a region within a network.

The ISDN subscriber number is a variable-length field containing the number used to reach the subscriber in the same local network or numbering area. The ISDN subaddress can be a maximum of 40 digits. This is not part of the ISDN numbering plan but provides additional addressing information.

TABLE 10.2   Call information phase messages.

0000 0001 Alerting	ALERT
0000 0010 Call Proceeding	CALL PROC
0000 0011 Progress	PROG
0000 0101 Setup	SETUP
0000 0111 Connect	CONN
0000 1101 Setup Acknowledge	SETUP ACK
0000 1111 Connect Acknowledge	CONN ACK
0010 0100 Hold	HOLD
0010 1000 Hold Acknowledge	HOLD ACK
0011 0000 Hold Reject	HOLD REJ
0011 0001 Retrieve	RET
0011 0011 Retrieve Acknowledge	RET ACK
0011 0111 Retrieve Reject	RET REJ
0100 0101 Disconnect	DISC
0100 1101 Release	REL
0101 1010 Release Complete	REL COMP
0110 1110 Notify	NOTIFY
0111 0101 Status Enquiry	STAT ENQ
0111 1011 Information	INFO
0111 1101 Status	STAT
1111 1011 Key Hold	KEY HOLD
1111 1100 Key Release	KEY REL
1111 1101 Key Setup	KEY SETUP
1111 1110 Key Setup Acknowledge	KEY SETUP ACK

10.5.3.1 Q.931 message applications.   There are four applications supported by Q.931 procedures. They are circuit-mode connection control, packet-mode access connection control, user-to-user signaling, and global call reference.

Circuit-mode connection control involves the set up, supervision, and subsequent release of B channel connections and

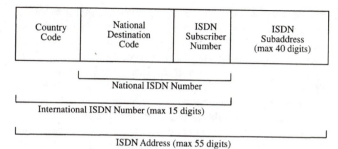

Figure 10.10 ISDN address format.

their resources for call control. This is mostly related to voice and data calls over a circuit-switched connection.

Packet-mode access connection control involves the set up, supervision, and release of data connections using packet-switched services over circuit-switched connections. This is an ISDN-specific feature.

User-to-user signaling consists of control messages sent from one user to another over the D channel. The users in this context are connection endpoints. While the signaling messages do not use B channels, they may well be associated with connections already established over B channels. This could be used by two PBXs to send signaling messages to one another over the PSTN. This is different from Q.931 signaling, which relates to ISDN circuits. User-to-user signaling is transparent to the ISDN and the SS7 network and is used solely by the user.

There are four classifications for Q.931 messages: call establishment, call information phase, call clearing, and miscellaneous messages. These messages are not used in the public network. The D channel is not extended from telephone office to telephone office; it is only used from the user network to the public network. Once the D channel message has been received by the local exchange, it is forwarded to SSP, which is an SS7 function. The ISDN signaling message is then changed to an SS7 message, which is then forwarded through the public network.

10.5.3.2   Call establishment messages.   The SETUP message is used to initiate a call connection. It can be sent in either direction (network to user or user to network). If the network is sending the SETUP message, it means that a user is requesting a connection, and the network is forwarding the SETUP message to the destination user (network to user). If the user is sending the SETUP message, obviously the user is requesting a connection to be established with another user.

A SETUP ACKnowledge is sent in response to a SETUP message. It indicates call establishment has begun but more information is needed. If more information is not needed, this message is not sent.

CALL PROCeeding indicates that call establishment has been initiated. It can be sent in either direction and is of local significance only.

ALERTing indicates that the called party is being alerted (the ISDN phone is ringing). This message can also be sent in either direction (user to network or network to user).

CONNect means the called terminal has accepted the connection request (the called party has answered). The B channel should now be connected, and conversation can begin once the message is acknowledged. This message is also sent in both directions.

CONN ACKnowledgment indicates that the call has been awarded to the user and is sent in response to a CONNect message. When this message is received by the user sending the CONNect message, conversation begins.

PROGress is used to report the progress of a call establishment. It can be sent in either direction during any part of the connection establishment phase (but only after a SETUP message has been sent).

10.5.3.3 Call information phase messages.

The call information phase messages used are listed in Table 10.2. These messages are sent after a connection has been established. They are sent in relation to connections on B channels where conversations are in progress. The RESume message is used to resume a call that has been previously suspended. It is only sent in the user-to-network direction. Calls can be suspended temporarily, allowing the calling or called party to initiate a connection to another entity. We will discuss suspension later on.

RESume ACKnowledge indicates that the resume request has been granted, and the call is reestablished. This message is only sent in the network-to-user direction.

TABLE 10.2   Call information phase messages.

Alerting
Call Proceeding
Progress
Setup
Connect
Setup Acknowledge
Connect Acknowledge

RESume REJect is sent if the call cannot be resumed. For example, a B channel connection may not be available to resume the call, resulting in the reject message. This is only sent in the network-to-user direction.

SUSpend allows a specified call to be temporarily suspended. The B channel is made available for other calls but not through Q.931 release procedures. The difference is that a RELease message flushes all buffers associated with the specified connection and the network no longer maintains called/calling party identities.

With the SUSpend message, the identity of the called and calling parties is maintained, and resources dedicated to the connection are maintained. This makes reestablishment of the connection much quicker (the call establishment phase is not used). The SUSpend message is only sent by the user to the network and is of local significance only.

SUSpend ACKnowledge is returned in response to a SUSpend message and indicates that the B channel has been released. Charging is also stopped at this point. The network sends this message back to the user.

SUSpend REJect is sent by the network to the user, indicating that the request to SUSpend a call has been rejected.

User information can be sent in either direction, but a user (not the network) always initiates it. One user sends user information to the user on the other end of a connection. The ISDN standards do not define these messages, although it is assumed that this would be call control information sent by two devices at user networks.

10.5.3.4 Call clearing messages. DISConnect is sent in either direction and is always initiated by either user. It is used to begin release procedures for a channel and all associated circuits. When received by the network, the network begins releasing circuits through the PSTN used for the connection.

RELease is sent in response to a DISConnect and indicates the channel(s) to be released. All associated circuits and resources reserved for the connection are released at this point.

RELease COMPlete is sent in response to a RELease message, indicating that all resources and associated circuits have been released. It should be noted here that this does not necessarily mean that circuits connected through the PSTN have also been released. Only the channel(s) between the user and the network are indicated in this message. SS7 messages then proceed through the PSTN to release circuits through the PSTN. Table 10.3 shows the call clearing messages.

10.5.3.5 Miscellaneous messages. Congestion control can be sent in either direction and is used to begin or end flow control procedures on messages associated with user information (end-to-end signaling messages).

FACility is used to request a supplementary service. It can be sent in either direction but is always initiated by a user.

INFOrmation provides additional signaling information during any of the call phases we have just discussed. It is initiated

TABLE 10.3   Call clearing messages.

Hold
Hold Retrieve
Hold Reject
Retrieve
Retrieve Acknowledge
Retrieve Reject

by either user and is sent in both directions (user to network or network to user).

NOTIFY is used to send information that pertains to an established call.

STATUS is used to send error information as well as the status of a call in progress. It is sent in response to a STATUS enquiry message. The STATUS enquiry message is sent to initiate a STATUS message (request status of a call in progress).

Table 10.4 shows the miscellaneous messages used.

10.5.3.6 Q.931 message parameters. A number of parameters are used with Q.931 bearer capability messages. These parameters are described in this section.

Bearer capability parameters provide detailed information needed to establish the desired services over a B channel. They are defined in I.231 and I.232. The protocol options at each layer are defined. The information is divided into four groupings: bearer services, access attributes, information transfer, and general attributes.

Bearer services define the access methods to network functions or facilities, how information is to be transferred over the network, and general attributes such as quality of service and supplementary services.

TABLE 10.4 Miscellaneous messages.

Disconnect
Release
Release Complete

Access attributes define the type of channel to be used as well as the user data rate. The signaling and information access protocols for layers 1, 2, and 3 are also identified.

Information transfer attributes define the bit rate for circuit-switched connections; whether circuit-switched, packet-switched, or Frame Relay services are to be provided for the call; and the call configuration (point-to-point, point-to-multipoint, or broadcast).

General attributes are used to identify supplementary services, quality of service, and interworking parameters. Supplementary services are defined as

- Number identification

- Call offering

- Call completion

- Multiparty

- Charging

- Community of interest

The following is a description of the parameters provided for bearer services.

Call identity.   Used with the SUSpend message to identify a suspended call. It is used as a reference so that when a RESume message is sent, it can be correlated with a call. The call identity is assigned by the initiator of a SUSpend and is assigned at the start of suspension (when the SUSpend message is sent).

Call state.   Identifies the state of a call. The state may be indicated as active, detached, or disconnect request.

Called/calling party number.   The subnetwork of both the called and the calling party. It is the subnetwork portion of the ISDN address. There is a separate field for the called party and one for the calling party.

Called/calling party subaddress.   The subaddress of the called and calling party. This is also part of the ISDN address. There is a separate field for the called party and one for the calling party.

Cause.   Provides diagnostic information in a number of messages. For example, in the RELease message, the cause parameter identifies the reason for call clearing. Cause codes are not necessarily specific. In the case of the RELease message, one possible value is "normal call clearing," which indicates someone hung up.

Channel identification.   Identifies the B channel for call establishment messages as well as call information phase messages.

Congestion level.   Originally intended to provide congestion levels, this parameter currently supports two values: receiver ready and receiver not ready. Eventually, this parameter may be expanded to provide individual levels of congestion, but this has not yet been defined.

Display.   Provides information in ASCII format to be displayed on an ISDN terminal (such as the display found on ISDN telephones).

Facility.   Indicates the invocation of supplementary services.

Feature activation. Used in the call establishment phase to activate specific features.

Feature indication. Provides status information to the user of supplementary services.

High-layer compatibility. Used in the SETUP message during the call establishment phase to identify the terminal type connected at the S/T reference point. This information is passed transparently through the network as user-to-user information.

Keypad. ASCII representation of characters entered on an ISDN terminal's keypad.

Low-layer compatibility. Provides information transfer capability and transfer rate and identifies protocols at layers 1 through 3. Used to check compatibility of lower layers in the network.

More data. Indicates that a USER INFORMATION message has been segmented and additional information is being sent in another USER INFORMATION message.

Network-specific facilities. Allows the specification of certain facilities that are unique to a network.

Notification indicator. Currently the only values supported are user suspended, user resumed, and bearer service charge. This information is sent in the NOTIFY message to provide information pertaining to a call but not the status of a call.

Progress indicator. Used in both call establishment and call clearing messages to indicate an event that took place while the call was in progress.

Repeat indicator.   Indicates that information elements have been repeated, and only one possibility should be selected.

Restart indicator.   Initiates a restart on a channel. A restart will flush all of the buffers and reset all associated counters and timers to zero.

Segmented message.   Indicates that the message received has been segmented and the rest of the message will be sent in a subsequent message.

Sending complete.   Used in the call establishment phase to indicate completion of the called party number. This is an optional parameter used in the SETUP message.

Signal.   Used during the call establishment and call clearing phases. Stimulus terminals (ISDN telephones) interpret this parameter and generate tones (such as ringback, busy, and dial tone).

Switchhook.   Tells the network the status of a stimulus terminals switchhook. The status can be either on hook or off hook.

Transit network selection.   Identifies a network or sequence of networks that should be used to complete a call. If a sequence of networks is identified, the parameter is repeated within a message.

User-to-user information.   Has no network significance and is transparent to the network nodes. Used to send user information end to end (between two private switches).

Figure 10.11 shows the sequence of events and the message exchange that takes place during the call establishment, call information, and the call clearing phases. The figure also shows the relationship between ISDN messages and SS7 messages. Remember that ISDN messages are not sent through the PSTN but are "converted" to SS7 messages and passed through the

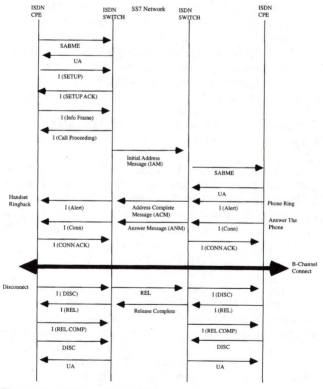

Figure 10.11   ISDN and SS7 call establishment/call clearing.

network as SS7 signaling messages. User-to-user information is passed transparently through SS7, using the pass-along or end-to-end method.

Notice that the SS7 messages must be sent independently from exchange to exchange. In other words, the IAM message is sent from one exchange to another but not any further. The two exchanges must then establish a connection between each other. In the meantime, the farthest exchange begins setting up another connection between itself and the next adjacent exchange by sending an IAM message. This process is repeated until a route is established end to end, and connections are reserved until the called party answers.

Packet-mode call establishment procedures are similar to those used for circuit mode. Subsets of the same messages are used.

ISDN does not address the needs of future networks. As we have seen, ISDN is sufficient for low-speed data, digitized voice, and slow-scan video. To support high-speed data, broadcast-quality video, and interactive multimedia applications, a new version of ISDN is needed.

10.6   BROADBAND ISDN ARCHITECTURE

There are two types of interfaces, or access points, defined in the broadband ISDN (BISDN) standards: the network-to-network interface (NNI) and the user-to-network interface (UNI). The NNI allows multiple networks to be interconnected, forming one large cohesive global network. The UNI allows subscribers to interconnect equipment from any vendor to the BISDN. This is not unlike the basis for ISDN.

The physical layer for BISDN must rely on fiber optics. Broadcast video requires high sustained data rates, which existing copper infrastructure cannot support. Deployment of SONET is a requirement to support data rates up to 600 Mbps.

ATM is the transport for BISDN. It uses cells (a cell is the equivalent of a packet, with some significant differences). One of the most significant differences between cell relay and packet switching lies in the header and the fixed cell size (compared to packet switching, which uses large headers and trailers and variable-sized packets). Packets are routed according to the destination address in the packet header, and cells are routed based on a connection, identified in a small header.

ATM is independent from the physical layer. This allows ATM services to be offered over SONET or existing copper-based services. Although twisted pair copper cable cannot support the speeds of some ATM services, lower speeds can be supported. ATM is really a multiplexing method that uses

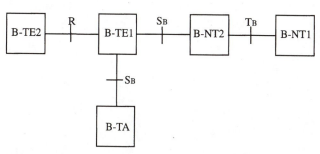

Figure 10.12 BISDN reference model.

nondedicated slots for transmission of cells, making it incompatible with channelized (T-1) services.

Another unique feature of ATM is that it can support variable transfer rates for the same connection, whereas channelized services must provide the same transfer rate throughout the duration of a connection. In other words, with ATM the transfer rate can be changed while a connection and data transfer are in progress, something channelized services cannot support.

Figure 10.12 shows the BISDN network. This model was derived from the narrowband ISDN (NISDN) model, using the same functions found in NISDN. The functions change somewhat, due to the nature of broadband transmission.

11

Frame Relay

Originally developed as an interim solution and stepping stone to ATM technology, Frame Relay has quickly become the first choice for many corporations faced with connecting their LANs to remote offices. As file transfer and remote access to large databases become more critical to businesses, Frame Relay is an excellent choice for many reasons.

Frame Relay is a layer 2 protocol. It encapsulates data (including protocol headers from other network technologies) into a frame and sends it over a Frame Relay network to its destination. If a message originates in a TCP/IP network, the TCP/IP header information is encapsulated into a Frame Relay frame and sent to its destination. If the destination is also a TCP/IP network, the header information can then be used to continue routing the data over the TCP/IP connection.

The unique feature of Frame Relay is that it does not provide any error detection/correction or real flow control (between devices). The only flow control provided in Frame Relay is used by the network to prevent network congestion. Even then the procedures are very simple, requiring little or no processing on the part of the network devices. That is the real intent of Frame Relay, to remove all of the requirements for processing within the network and let the upper-layer protocols (such as TCP/IP or ISDN) provide those procedures. This means the end devices must provide the flow control procedures and error detection/correction outside of the network. The end result is a very efficient network protocol that introduces very little delay because there is so little processing required to deliver messages.

This technology was developed for bearer services over ISDN circuits. The intent was to provide a streamlined transport protocol for data transfer, eliminating many of the procedures found in packet-switching networks such as X.25. The ITU, ANSI, and the Frame Relay Forum developed the Frame Relay standards. The ITU continues development of international Frame Relay standards, while the ANSI continues its development of Frame Relay standards for use in the United States. Frame Relay was developed from the X.25 standards and uses protocol procedures that are similar to those found in X.25 and ISDN.

Frame Relay is based on packet switching. Time slots are used whenever data needs to be sent but are not dedicated to any one connection (as is the case in circuit-switched services). Frame Relay is different from cell relay in at least one aspect: Frame Relay supports variable-sized frames, whereas cell relay uses fixed-sized cells (they are the same as a packet, just different terminology).

The Frame Relay standards are documented in ITU I Series Recommendations and are broken up into different descriptions. The ITU I.233 (and the ANSI T1.606) defines the services provided by Frame Relay, as well as the overall functions in

Frame Relay protocols. ITU I.370 defines the procedures used in Frame Relay to manage network congestion. The functions of access signaling and data link control are defined in ITU Q.922, "Core Aspects." ITU Q.933 defines the protocol for signaling information used to establish and maintain virtual connections over the Frame Relay network. ITU Q.922 defines the protocol used for end-to-end delivery of data as an option. ANSI has not defined this option in any of its standards.

There are two different data rates defined in Frame Relay, the committed information rate (CIR) and the access rate. CIR is the rate at which the network has agreed to carry data and is different from the access rate. The access rate is the data rate provided at the physical connection at either end of a Frame Relay virtual connection.

The CIR is enforced through rate enforcement. Frames that exceed the CIR will be carried over the network only if there is available bandwidth. If sufficient bandwidth is not available, the frame is discarded. Any frames with the discard eligibility bit automatically set, and found to exceed the CIR, are automatically discarded regardless of the available bandwidth.

Another parameter that must be negotiated at connection establishment is the committed burst size. This is the maximum number of bits per second the network will transfer during any measurement interval.

Frame Relay was developed for data transfer and was never intended for voice transmission. However, due to the slow deployment of ATM, many companies have found Frame Relay suitable for voice transfer. In fact, the Frame Relay Forum is actively defining standards to support voice over Frame Relay networks. The savings can be 25 to 35 percent over conventional T-carrier facilities.

As with T-carrier (T-1, T-3, etc.), the voice is converted first into digital form (using PCM). It is then compressed using Digital Speech Interpolation (which deletes any pauses or hesitations in the audio) and sent over the Frame Relay network with as few hops as possible (decreasing delay). Currently, voice over Frame Relay is limited to one network. Voice cannot be sent between two Frame Relay networks under today's Frame Relay standards.

Another limitation is that Frame Relay cannot support prioritized traffic. Voice traffic is treated with the same priority as data traffic. When the two are mixed, voice does not take precedence over data, which could cause additional delays if the traffic mix got heavy. Because Frame Relay allows for variable packet sizes, data packets can be up to 100 megabytes long, causing delays in the delivery of smaller voice packets. This may change as new standards evolve.

A Frame Relay access device (FRAD) can be used to segment large data packets into smaller packets, reducing the potential for delays. A FRAD is a device that sits on the edge of the network, acting as an interface between the voice switch and the Frame Relay network. FRADs can be used for a variety of applications, such as accepting traffic from non-Frame Relay networks and encapsulating the data into Frame Relay packets for transmission over the Frame Relay network.

One such device even allows SS7 traffic to be transferred over a telephone company's Frame Relay network, reducing the cost of SS7 facilities. In the case of this product (made by Tekelec), the FRAD actually discards extraneous FISUs, reducing the amount of traffic sent over the Frame Relay network. This is where the cost savings come in, because it reduces the amount of traffic required between SS7 nodes. The

SS7-FRAD (patent pending) performs other functions as well, but this gives you the principal idea behind it. FRAD provides interworking between different network types and Frame Relay networks.

A FRAD can also prioritize traffic before sending it over the Frame Relay network. The FRAD does not actually add priority information, but it receives traffic from other protocols, uses the procedures of those protocols to apply prioritization, and buffers lower-priority traffic while sending high-priority traffic. Some FRADs also provide voice compression, reducing the bandwidth requirements from 64 to 4 kbps.

FRADs are currently available to support ATM, TCP/IP, SS7, and other protocols. They are point-to-point devices, requiring a FRAD at both ends of the Frame Relay connection. This is analogous to a modem, which requires another modem at the other end of the connection.

The advantage of using Frame Relay is its speed and low overhead. Most protocols require a lot of processing at each of the network nodes. Frame Relay requires very little network node processing, relying instead on the upper protocols and applications to provide management procedures within the end devices. This, of course, reduces the delay time in the network and allows packets to be sent much more quickly (which means more throughput).

Figure 11.1 shows a Frame Relay packet and its fields. As you can see, there is very little to Frame Relay. The addressing used by Frame Relay is the same concept used in ISDN, using a DLCI to identify the connection.

The forward explicit congestion notification (FECN) and backward explicit congestion notification (BECN) fields are used by Frame Relay nodes in the network to notify other network

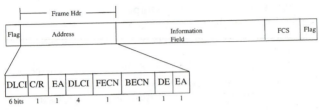

Figure 11.1   Frame Relay format.

nodes of congestion. When congestion conditions are encountered, Frame Relay makes no attempt to manage lost frames. The Frame Relay nodes simply discard the frames and rely on the upper layers to keep track of lost frames.

Congestion notification cannot be sent to the source of the congestion. Only the nodes that are adjacent to the congested node act on the FECN and BECN, slowing transmission to the congested node until the congestion subsides. Most carriers do not implement congestion control using the FECN and BECN because of this limitation. Instead, many use Consolidated Link Layer Management (CLLM) at the end devices because this protocol enables end nodes to send congestion and status information to other nodes in the network. CLLM is a management protocol, sent in the payload portion of Frame Relay frames.

Nodes in the network do not provide layer 3 processing and provide very little layer 2 processing (such as error correction and flow control). Instead, Frame Relay follows a very simple rule: If the data is not in error, it is routed to the destination. If the data is found to be in error (or the frame is in error), the frame is discarded. The end nodes are then responsible for retransmission and end-to-end flow control. This enables network nodes to process frames much more quickly and more efficiently.

Originally, Frame Relay was designed for data transfers with large throughput requirements. Applications sending small-sized frames in bursty form are not good candidates for Frame Relay (it will work, but it is not as cost effective).

Frame Relay is not dependent on ISDN. There are many Frame Relay networks in use today that are running independently of ISDN. The Frame Relay protocol was developed to provide packet-switching services for ISDN subscribers without having to use X.25. The X.25 protocol was designed for use on circuits that were not as reliable as today's, and it has a lot of overhead and procedures requiring processing by the network nodes. The idea behind Frame Relay was to eliminate all of the processing requirements in the network and to move the recovery procedures out to the end nodes. This is a good idea, as long as the circuits are reliable and there is not a lot of noise on the line. If the line is noisy and data is corrupted as a result, Frame Relay is no longer an efficient means of data transport because the upper layers will require retransmission too often, flooding the network with retransmitted frames.

12

Cellular Networks

Cellular networks were developed as a more efficient methodology for mobile communications. The early radio-telephone
networks consisted of a limited number of radio frequencies
that had to be shared among hundreds of subscribers. The network architecture was such that each of these frequencies provided coverage for many miles. This means only a limited
number of subscribers can be supported at the same time,
which is why radio telephones were never very popular. Users
experienced long wait times before they could place calls. Cellular architecture resolved this problem by reengineering the
network topology and the usage of frequencies.

If the coverage area is made smaller, and the frequencies
divided into smaller geographic areas, more subscribers can be
supported. The available frequencies can be reused in other
cells, as long as the cells are not adjacent to one another. This
allows networks to use the existing spectrum, reusing the same
frequencies within their network.

This concept is known as frequency reuse. If the available frequency spectrum is divided into blocks of frequencies, those blocks can then be allocated into cells. Each cell then uses a different range of frequencies. If allocated properly, the same frequency would never be assigned to two adjacent cells, and there would be sufficient distance between cells with the same frequencies that they would not interfere with one another.

If the cells covered a small geographic area, the transceivers used would not require as much power. The less power used to transmit, the less interference with neighboring cells. In today's cellular networks, frequencies are divided into seven blocks and are assigned to cells accordingly. Figure 12.1 illustrates this structure.

A cell can cover up to a 10-mile radius in any kind of pattern. Antenna technology allows antenna arrays to be arranged to provide any type of coverage needed for a geographical area. Directional antennas and new smart antennas have enhanced the control given to cell sites. This provides better-quality audio as well as more efficient use of the frequency spectrum.

This is the basic architecture of modern-day cellular networks. How can we possibly improve it? If the cells could be made even smaller, the transceivers would require less power. If transceivers required less power, cellular phones would require less power as well. Their batteries would last longer and could weigh less, meaning that the phones could be made smaller. The frequency reuse pattern could be repeated many more times, providing support for many more subscribers. This is the concept being used today for new personal communication services (PCS) networks.

PCS is not a technology but a level of services. There is not much difference in the operation of a PCS network and a modern

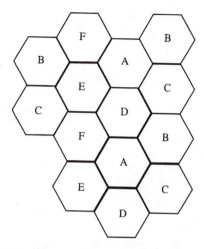

Figure 12.1 Cell structure and frequency reuse.

day cellular network, apart from the higher frequencies allocated for PCS networks. Both are digital, both use the same types of entities, and both rely on intelligent networking to access databases and communicate with other networks. It's the type of services provided, such as one number access for your cellular and home phone, voice mail, and pager.

12.1 DIGITAL VERSUS ANALOG

Within the United States, there is only one kind of analog network. This is referred to as the Advanced Mobile Telephone System (AMPS) and is presently deployed throughout North America. Analog systems are quickly being replaced with digital systems. The analog system has a limit to the number of subscribers it can accommodate. Each frequency is capable of handling one call. There is a limit to how many frequencies there can be in an area (or cell) at the same time, which places a limit on how many subscribers can be supported within a cell at one time. This is one of the reasons the analog cellular system is being replaced with a digital one.

Two technologies are currently being investigated by cellular network providers looking to convert to digital. Time division multiple access (TDMA) was introduced in 1992 and is deployed in several networks throughout the United States. It is an air interface technique also used in the Global System for Mobile Communications (GSM) networks in Europe.

Code division multiple access (CDMA) is a more recent solution and is quickly gaining popularity for a number of reasons. CDMA is being touted as capable of handling more transmissions per frequency, higher data rates, and more secure transmission than TDMA. CDMA also provides additional control of power levels when transmitting, offering extended battery life to cellular telephones and less interference between cells. There are several differences in the way TDMA and CDMA operate.

GSM is a European technology, now being deployed in U.S. networks. GSM development started in 1982, but it was not until 1992 that GSM was first deployed in Germany. GSM is a standard defining the entire network, not just the air interface. It addresses signaling, control of different entities within the network, and network management.

Cordless telephone technology works in similar fashion to cellular but on a much smaller scale. A base station must be used within close proximity (creating a "pico" cell) to the handset and provide an air interface. The air interface can be analog or digital, depending on the standard used.

In the rest of the world, there are several choices for cordless telephone technology, most based on Cordless Telephony 2, or CT2. This is a second-generation standard defining the air interface for wireless communications at pedestrian speeds, and it qualifies the standard for use in PBX systems as well as cordless telephones.

CT2 originated in the United Kingdom in 1992 and has been adopted by the European Telecommunications Standards Institute (ETSI) as that nation's cordless telephone standard. Several other standards derived from CT2 are used in Europe and the rest of the world. These are discussed in the next section. CT2 is also used in the United States.

There are several new U.S. standards under development that are derived from CT2 and other standards. These are listed below:

- Personal communications interface (PCI) (based on CT2)

- Wireless customer premises equipment (WCPE) (based on the Digital European Cordless Telecommunications standard, or DECT)

- Personal access communications system (PACS) (based on the Public Handphone System, or PHS, and WACS)

- Wireless access communication service (WACS) (developed by Bellcore)

Some of these cordless telephone standards may become applicable to PCS solutions. For example, one company on the East Coast currently offers a service using both cordless and cellular telephone technology. The user gets a cordless telephone (based on CT2) that works as a portable telephone around the home. However, as soon as the user roams outside of the base station's range, the cellular network picks up the signal. The handset is a dual-mode handset capable of working in both CT2 and AMPS mode. The service has been fairly popular, although it is currently limited to specific service areas.

International cellular networks are far from compatible with those in North America. The original intention of the ITU was to create a universal network, allowing subscribers to seamlessly roam from network to network, country to country, using their cellular telephones. Unfortunately, given the solutions under development today, and the choices of both U.S. and international cellular providers, it is unlikely we will ever see such a network.

Europeans do enjoy seamless roaming in European countries, but outside of Europe, countries have chosen their own standards. Japan, for example, uses a standard proprietary to that country; it is not used anywhere else. The Middle East uses a variety of different technologies as well, making seamless roaming from the Middle East to Europe impossible. Of course, the United States has followed its own standards, making U.S. cellular telephones useless overseas.

Much of this difference lies in the frequencies allocated to cellular use. The Federal Communications Commission (FCC) has no jurisdiction in other countries and does not influence how frequencies are allocated outside of the United States. Different countries have different needs and must follow their own requirements for frequency allocation.

Europe's cellular industry is unregulated, with competition encouraged. There must be at least two cellular providers in each region, which is different than in the United States, where only two service providers are allowed per area. This, too, is under change after the passing of the Telecommunications Act of 1996. Now, wireline companies are competing for wireless subscribers, and wireless companies are looking at the local telephone service market.

Some analog cellular systems are still used internationally, although most of the networks have been converted to digital. The international cellular industry has grown much more quickly than the U.S. market because many countries do not have an existing advanced wireline communications infrastructure, forcing them to develop either a wireline network (at considerable cost) or deploy a cellular one. In many regions, it is far more cost effective to provide cellular services than try to maintain wireline infrastructure. The two international analog systems are Nordic Mobile Telephone (NMT) and Total Access Communication Systems (TACS). They are used mostly outside of Europe.

The most widely used digital cellular standard in the global community is GSM. This is actually a European standard, but it has been adopted for use in many other regions of the world, including parts of the United States. GSM provides a fully digital cellular solution. The air interface used for GSM is TDMA. There are some differences between the TDMA used in the United States and the version used in Europe. Most of these differences lie in the frequencies used and the number of transmissions that are multiplexed onto a signal channel. GSM is capable of supporting eight transmissions per channel, whereas U.S. networks support three transmissions per channel. TDMA can support more than this, depending on a number of factors, such as channel spacing. The United States uses 30-Hz channel spacing, and Europe uses 200-Hz channel spacing, affecting the amount of bandwidth available.

Many cordless telephone standards are used throughout the world. In Europe, the DECT standard is widely used. This was derived from CT2 and adopted by ETSI as a standard. Japan uses its own standard, PHS, which is not compatible with any other cordless telephone standard.

Before discussing the various cellular solutions, we need to understand how cellular systems work and how the network elements perform. There are differences in the various technologies, but for the most part, there are more similarities. We will first discuss how cellular networks operate and then look at the differences between the various solutions.

12.2   CELLULAR OPERATIONS

To understand cellular technology, you must first understand how a cellular network operates and the entities used within the cellular network. Figure 12.2 is the basic model for a cellular network. Some technologies, such as GSM, add more entities, providing different features to the network.

Cellular networks are deployed according to markets. The wireline industry deploys their services according to service areas, or LATAs. These were defined by the courts according to demographic data. In the wireless industry, the areas are called metropolitan statistical areas (MSAs) and rural statistical areas (RSAs). There are some 305 MSAs and 482 RSAs as of this writing.

New PCS networks have been divided differently than cellular networks. Metropolitan trading areas (MTAs) and rural trading areas (RTAs) are defined for PCS networks and do not cover the same territories as MSAs and RSAs.

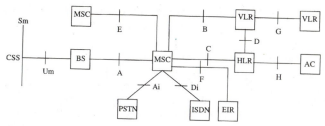

Figure 12.2 Cellular network model.

Two carriers are allowed to provide cellular services within each MSA or RSA. Frequencies are allocated to both of the carriers in blocks. The two blocks of frequencies within a market area are labeled as system A or system B. Cellular phones must be able to work in both systems, regardless of the technology used in that network. For example, if system A is an analog AMPS network, and system B is a TDMA IS-54 network, subscribers must be able to use their phones on both systems.

There are 21 setup, or control, channels within each frequency block. They operate in two directions, forward and reverse. The forward channel is used to page the mobile unit, whereas the reverse channel is used by the mobile unit to request a call setup.

When a cellular phone is powered on, it will search the 21 control channels for the strongest signal. Each control channel is associated with a cell site. When the cellular phone determines which signal is the strongest, it sends a call request to the cell site over the control channel. The cell site will then determine which voice channel is available and send a signal to the cellular telephone identifying the frequency. The cellular telephone can then begin the connection process on the voice frequency.

The heart of the cellular network is the MSC, which manages the routing of calls within the cellular network. It also controls handoffs between cell sites, access to system features, and access to network databases. There is typically an MSC (Mobile Switching Center) per MSA/RSA. The MSC also coordinates handoffs between cell sites, manages mobile paging (the process used to alert a cellular phone), processes registrations from cellular phones when they are powered up, and provides and/or manages connections to the PSTN.

The MSC provides access to databases as well. All cellular networks have an HLR. This database provides information about all subscribers in that home area. The cellular company providing service in that MSA/RSA owns the HLR. The subscriber's cellular phone is encoded with an equipment serial number, an MIN, and a cellular telephone number. This information is stored in the HLR.

If you have owned a cellular telephone, you know you had to choose the area you wanted to be your home area. Any calls made outside of this area use roaming services, which may be provided by the same cellular company or by a competitor. The data regarding your telephone and the services you pay for are kept in the database located in that home area. When you roam, the cellular company servicing your call must access the HLR in your home area to determine who you are and how to handle service to your phone.

All MSA/RSAs have access to a VLR as well. This is also a location database, but it provides a little more detail than the HLR regarding your location. The HLR only knows the last

MSC servicing your telephone. This information is maintained until updated by a VLR. The VLR knows the last cell site to provide you service. Here is how it works:

When a call is directed to your cellular phone, the telephone company handling the call routes the caller to your home MSC. The MSC will then query its HLR to see where you were last located (which MSC you last made a call from). If the HLR says you were in a different network, the MSC forwards the call to that MSC.

When the call arrives at the remote MSC, the MSC will query its HLR. When it does not find your MIN in the database, it looks in the VLR. The VLR will identify whether or not your phone is active in that network, and if so, which cell site is currently servicing your phone. This information is kept up to date; however, the VLR database will remove the record after a predetermined period if there has been no activity on your cellular phone. While your cellular phone is idle, it periodically sends out a signal to the cell site to let it know the phone is still active. This signal will include the MIN assigned to the phone so that the information can be stored in the VLR for that area. If you move to another network, serviced by a different MSC, your home HLR is updated to reflect the new MSC. It is the responsibility of the VLR to provide this updated information.

The VLR is a very dynamic database. Data is not stored forever because this would quickly fill the database beyond capacity. Instead, if a record ages beyond a specified time period without updates, the record is deleted.

The data in the HLR consists of the last VLR to provide registration information regarding the cellular phone. It provides the SS7 point code and subsystem number of the VLR (both addresses used in the SS7 protocol), as well as the subscriber data such as MIN, equipment serial number (ESN), and mobile state (active/inactive).

The cell site consists of two parts, the base station controller (BSC) and the base station transceiver (BTS). The BTS is the radio transceiver and the antennas used at a cell site. The combination of antennas and transceivers is called the base station subsystem (BSS). The antennas are connected to a system that allows the antennas to be switched to different transceivers. This is controlled by the BSC.

Since there are several components in the BSS, we will talk about them one at a time. First, we will talk about the various antenna systems used. Antennas can provide coverage in numerous ways. Omnidirectional antennas provide coverage in a uniform circular pattern.

Directional antennas provide coverage in one specific direction. These antennas are good in areas where signals need to be sent in a specific direction, in cities where buildings prevent a uniform circular pattern, or against hills and mountainsides where omnidirectional patterns would be significantly blocked on one side. Directional antennas can also allow transceivers to operate at lower power levels because the signal is more concentrated. This is only an advantage in systems such as CDMA, where power levels can be fluctuated depending on signal strength.

Sectorized antennas are divided into three to six different sections. Each section emits a signal, in varying degrees of coverage. This provides wide coverage and allows signals to be concentrated in each sector rather than spread over an omnidirectional pattern.

Smart antennas use sectorization techniques but further divide the sector into single signal beams. The signal strength of cellular phones is monitored, and the beam receiving the strongest signal is used to service the cellular phone. These systems use switched beam technology, which allows the antenna controller to switch from beam to beam. This provides the function of a directional antenna with the coverage of an omnidirectional antenna.

Another advantage to this type of antenna technology is that it provides stronger signals with less possibility of interference. This is the same as directional antennas that are able to concentrate their signals in one direction, without interfering with surrounding signals. For small coverage areas (microcells and smaller), a new technology called remote antenna driver (RAD) and remote antenna signal processing (RASP) is being used. A RAD is placed in the network, on top of street lamps and telephone poles. This small antenna communicates with a RASP over cable television facilities. The RASP is placed in operations centers and translates signals from cellular phones for transmission over the wireless networks. This is an attractive solution for cable television operators who are looking to move into wireless communications using the existing infrastructure. It can also be used to deliver cable television signals to homes without cable using wireless receivers in the home.

Multisubscriber units can be installed in apartment buildings or campus environments. These units allow standard two- and four-wire equipment to be connected via the cellular networks. All users are then multiplexed over the single air interface (acting as a concentrator). Single subscriber concentrators can be used to attach multiple telephone extensions to one air interface. This is how telephone companies in Mexico are delivering telephone service to residents where no telephone infrastructure exists.

Using cellular networks to deliver residential telephone service has demonstrated savings of 36 percent over wireline services. PCS may use the same techniques to provide wireless services to residential areas.

The cell site antenna system is attached to the base station transceivers. These transceivers are used to transmit and receive calls over the air interface. Each transceiver can support one or more multiple transmissions per frequency depending on the technology used over the air interface. In the United States, the TDMA solution supports three transmissions per frequency, whereas the same air interface (TDMA) in Europe supports eight transmissions per frequency (GSM).

The BSC is responsible for call processing and handoff procedures under the direction of the MSC. The BSC also provides audio compression and decompression. The handoff procedure is a complex procedure used when a cellular phone moves from one cell to the next. In CDMA networks, a soft handoff is used, where both the old and the new cell site handle the call at the same time.

The BSS is connected to the MSC. One MSC controls many BSSs. Remember that there is usually one MSC per MSA/RSA (in most situations). Now that we understand all of the components of a cellular network, we need to understand how they all interconnect and communicate with one another. Figure 12.3 shows the relationships of all of these components and how they are interconnected.

The air interface is between the BTS and the cellular phone (or handset or cellular telephone or whatever you want to call it). The air interface can be analog or digital and can use a variety of technologies to support it. We will be looking at the differences between these various solutions in a moment, but for

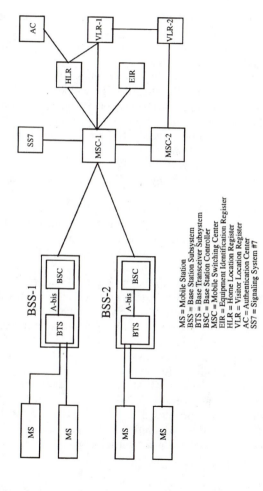

MS = Mobile Station
BSS = Base Station Subsystem
BTS = Base Transceiver Subsystem
BSC = Base Station Controller
MSC = Mobile Switching Center
EIR = Equipment Identification Register
HLR = Home Location Register
VLR = Visitor Location Register
AC = Authentication Center
SS7 = Signaling System #7

Figure 12.3 Cellular network model.

now know that there are differences between them and that they are not always compatible. Cellular phones are usually manufactured as *dual mode*, meaning they can operate over an analog air interface or a digital one.

From the BTS, there is a communications link to the BCS, which is the controller. The BCS can be collocated with the BTS or it can stand alone in the network, controlling several BTSs. The interface between them will depend on the strategy deployed. In a GSM network, the SS7 network provides services for the application protocol Base Station Subsystem Mobile Application Part (BSSMAP). In U.S. networks, there is probably an X.25 link using SS7 TCAP and IS-41 signaling protocol.

From the BCS, a link to the MSC provides channels for both voice and signaling data. The signaling data is usually run over an X.25 or SS7 link, with IS-41 protocol supporting the cellular signaling application. The voice link may be an ISDN line, as is the case in many GSM networks.

In the United States, the signaling application used is IS-41. This protocol runs on SS7 TCAP and defines the messages used to control handoffs and channel assignment. IS-41 does not define the air interface, and it works independently of the rest of the network. The air interface is defined by other standards such as TDMA or CDMA (depending on which solution is chosen).

In European networks, GSM defines the entire network. This includes the air interface (TDMA) as well as the links between all of the components. This is why many PCS vendors favor GSM over other U.S. solutions. GSM provides a standard for the entire network, not just a portion of it.

12.2.1 Call setup

The first and most important step to being able to make and receive cellular calls is registering with the network. When a cellular telephone is powered up, a signal is sent from the cellular telephone to the cellular network. This signal provides registration information, which is stored in the home HLR and VLR and, if the cellular telephone is in another network, in the visiting VLR as well.

The registration is sent to the MSC, which manages the registration of all cellular phones in its network. The MSC examines the MIN to determine whether or not the cellular phone should receive service. The MSC forwards the message to the VLR. The VLR updates an existing record if one exists. If there is no existing record for this MIN (remember the VLR is dynamic), a record is created. The VLR notifies the home HLR and requests a service profile to be used for the new record. The HLR sends the profile after authentication has been completed (a check to make sure there are no flags on record to deny service to the specified MIN).

There is always a chance that the cellular phone is registered in another VLR somewhere. The HLR knows the last serving MSC and can determine if a record exists in another VLR somewhere for the specific MIN. In this case, the home HLR must send a message to that VLR, instructing it to flush the record from its database.

When you dial a cellular telephone number, the office code of that number identifies the MSC that is registered as the "home" MSC for the subscriber. The subscriber may or may not be in that network, and the MSC must determine how to route the call. When the MSC receives the call, it examines the called number and queries its HLR. The HLR will identify the last MSC to serve the cellular phone. If the last MSC was the home MSC, the MSC can query the VLR to determine exactly which cell the cellular phone is now in. If it is registered in another MSC, the home MSC must transfer the call to the serving MSC. Before the transfer can take place, the home VLR must determine how to route the call to the now-serving MSC. The home VLR queries the serving MSC to determine how to connect the call and receives a temporary local directory number (TLDN) from the serving MSC. This TLDN is entered into the VLR, which will then update the HLR for future calls. The home HLR then sends the TLDN to the home MSC, which forwards the call to the TLDN.

The VLR in the visited area identifies which cell is serving the cellular phone and determines whether or not the cellular phone is active or inactive. If it is active, the MSC sends a signal out to the BSC requesting that the cellular phone be paged. The BSC will order the BTS to send a paging signal out to the cellular phone on the control channel for that cell. The paging signal will tell the cellular phone which frequency to use to receive the call.

When the cellular phone receives the paging signal, it switches to the proper frequency and sends confirmation to the BTS, which in turn sends confirmation to the BSC. The call can now be routed from the MSC all the way through to the cellular phone. All of this does take time, longer than wireline services would. You will notice a delay in the call setup when you dial a cellular telephone number, especially if the person you are calling is out of his or her home area.

12.3   CELLULAR NETWORK ARCHITECTURE AND PROTOCOLS

There are many standards defining wireless communications. Cellular systems, cordless telephones, wireless PBXs, and LANs are all defined under the classification of wireless communications. In this book, we will focus on the cellular market and data services offered through cellular providers, but we need to first understand the standards used in the cellular industry and the concept of "mobility." Cellular networks address three issues:

1. Terminal mobility
2. Personal mobility
3. Service portability

Terminal mobility allows a cellular subscriber to move from one network to another network or from one cell to another, without losing the ability to send and receive cellular calls. This is often referred to as *roaming*. In early cellular networks, subscribers had to obtain a roaming number and arrange for calls to be forwarded to that number. In today's networks, roaming is seamless, allowing subscribers to move around without special arrangements with the cellular providers.

The problem with terminal mobility is that it only addresses the mobility of the cellular terminal, and not the subscriber. Personal mobility allows the subscriber to move around without being tied to a particular handset or terminal. This is easiest achieved with wireless PBX, where an extension number is assigned to a user, and a special transponder worn by the user gives the PBX the location. It still requires some sort of device worn by the user to keep the network informed about the user's location. An alternative to the transponder is a PIN entered by the user at any phone. This tells the PBX where calls for that user should be routed.

Personal mobility is an issue under study and an objective of the newer PCS networks. Service portability is now under investigation. Part of service mobility is maintaining the same services you have at home in other networks. Number portability is part of service mobility and now is mandated by the FCC. In July of 1996, the FCC issued an order that all telephone companies in the top 100 MSAs must provide local number portability (LNP) by 1997. Cellular providers have been successful in delaying this requirement for several years and continue to fight the government.

Local number portability allows subscribers to keep their telephone numbers even when they change carriers. The first phase of LNP will allow subscribers to keep their numbers if they change telephone companies or service type but only within their rate center. The next phase of LNP is location portability, which will allow subscribers to keep their telephone numbers (including area code) when they move outside their rate center (a rate center is the local calling area in which subscribers reside). This is far more complex and is not part of the FCC mandate.

Telephone numbers for both wireline and wireless services are assigned to specific areas, determined mostly by geographical location. For example, most people know that the area code 212 is assigned to Manhattan. The office codes are assigned to specific central offices in that area code. With the new FCC mandate, a user who moves from Manhattan to North Carolina can keep the 212 telephone number. This is actually easier for wireless networks than it is for wireline network operators. The wireline network does not use databases to determine the location of a subscriber for incoming calls. With local number portability, database lookups will be required for all calls within an area code/office code (NPA/Nxx) if even one number is "ported."

Already, cellular networks search databases to learn the location of a cellular subscriber. Number portability will require an additional database search before a call can be connected, possibly extending the call setup time. The FCC already has mandates regarding call setup time for 800 numbers, and it is very likely that requirements will be set for all call setups as number portability starts to unravel.

Service portability can be achieved through the use of smart cards, intelligent networking, or both. Network databases containing subscriber information can be made accessible to all cellular providers through SS7. These databases are in use today for roaming and location management.

Smart cards are used throughout Europe (as part of the GSM network). These cards, called subscriber identity modules (SIMs), contain subscriber information and are inserted into GSM telephones through a slot on the bottom of the telephone. They are usually about the size of a credit card and can be used in any GSM phone. This is also being used in some PCS networks here in the United States. The advantage to using this type of system is that users can buy or rent cellular telephones anywhere, without activation. The phone becomes activated when the card is inserted into the phone, providing all of the information the network needs to perform authentication.

12.3.1 Time Division Multiple Access (TDMA)

TDMA first became available in 1992. TDMA is a digital air interface technology that allows cellular network operators to multiplex multiple transmissions over one radio frequency. This provides them with increased subscriber support using the available frequency spectrum. Estimates of deployment show some 2.5 million subscribers in 22 different countries currently use TDMA-compatible cellular phones.

In the United States, TDMA-compatible cellular phones must also be AMPS compatible so that the same cellular phone can be used in both AMPS and TDMA networks. This is known as dual mode (analog and digital). These phones are capable of detecting whether or not they are in an analog network, in which case they transmit analog signals, or a digital network, in which case they transmit in digital mode.

TDMA currently supports three digital transmissions over one frequency. This is a big boost to existing networks in dense subscriber areas as they struggle to support all of their subscribers over the already limited frequency spectrum. The TDMA standard defines support for up to 10 transmissions per frequency, but this has not yet been proven in any network.

In GSM networks, TDMA is divided into eight time slots rather than three. This is why GSM can support more subscribers per channel than U.S. networks. The difference lies in the channel spacing. In U.S. networks, AMPS networks use 30-kHz channel spacing, whereas in GSM networks, 200-kHz channel spacing is used. In the United States, it is important to follow the 30-kHz channel spacing for interworking between digital TDMA and analog AMPS networks.

When a cellular phone requests service from a cell site, the cell site will identify the frequency on which the phone should be transmitting (remember our earlier discussion about cellular operations) as well as the time slot to be used. This time slot is then assigned to the cellular phone and is not used by any other cellular telephone in the cell. The phone does not transmit anything until its time slot becomes available (usually every 4.615 ms). The phone is idle the rest of the time. This is known as bursty transmission. There is a big difference in frequency efficiency between TDMA and CDMA, discussed later in this chapter.

The standard that defines TDMA was developed by the EIA/TIA. Known as IS-54, this standard defines the air interface procedures. This includes the method used for accessing a cell site transceiver and digitizing the voice transmission. A newer standard, IS-136, is compatible with IS-54 and provides a number of enhancements over IS-54 for PCS networks.

The new IS-136 standard supports macrocellular service for large geographical coverage with low subscriber density. It also provides procedures for in-building networks, including "fixed wireless" networks. Fixed wireless networks provide cellular network support using fixed telephones and data terminals instead of portable handsets. These are popular where cabling is difficult or impossible.

IS-136 also supports seamless interworking with existing AMPS networks. This is critical to the success of PCS networks since they operate at different frequencies than AMPS networks. Cellular phones used for PCS networks operate in the 2-gigahertz (GHz) range, whereas AMPS networks operate in the 900-megahertz (MHz) range. This means dual-mode phones must be able to operate in both frequency ranges to support roaming between AMPS and PCS networks.

Some new features supported in IS-136 and not found in IS-54 include a message waiting indicator in the handset, calling number identification, alphanumeric paging capability (short messaging service), and sleep mode. The sleep mode automatically shuts down the phone when idle but periodically reactivates to check for messages.

It should also be mentioned that TDMA is used in GSM networks as well. If TDMA is so widely used, why do not all network providers select TDMA as their air interface? One of the fundamental problems in debate today is cellular telephone interference. Cellular telephones are notorious for interfering with aircraft navigational equipment and hearing aids, and although this is yet to be substantiated, there are reports of TDMA interfering with pacemakers.

The problem lies in the electronics required to support TDM. Transmissions must be multiplexed into time slots. This requires a timing system. Electronic clocks are run by crystals, which oscillate at specific frequencies. This oscillation emits radio frequency interference (RFI), which can interfere with the operation of other electronic equipment. To understand the effects of RFI, try placing your AM receiver antenna next to a fluorescent light bulb.

In CDMA phones, oscillators are not needed (unless they are working in dual mode, in which case an oscillator is needed). This means that CDMA cellular phones are less likely to interfere with other electronic equipment. There are a number of studies under way to determine exactly what effects there are from operating cellular telephones. None of the reports have been validated, and although there is a lot of concern over the impact on health and safety, there is no proof that cellular telephones present any health hazards, regardless of the technology used.

12.3.2 Coded Division Multiple Access (CDMA)

The EIA/TIA adopted the CDMA IS-95 standard from Qual-comm, a company that manufactures CDMA cellular equipment. There are many differences between CDMA and TDMA. The biggest one is in the use of the frequency spectrum.

In CDMA, several transmissions are sent over the same frequency without multiplexing. Instead, a unique digital code is added to the digitized speech for each transmission. Handsets receive all of the transmissions being sent over one channel and use microprocessors to decode each transmission and find the correct code. All other transmissions are then ignored.

Although this function requires additional processing power in the cellular telephones, the cost of microprocessors has decreased, making CDMA affordable as a cellular solution. The technique is not actually new; it has been used for military radio transmission for years. The method of spreading transmissions over the entire frequency spectrum is known as *spread spectrum technology.*

There are a number of advantages to CDMA. The IS-95 standard defines procedures for a complex power control method, designed to save on battery life and help prevent cochannel interference. As a CDMA phone is transmitting, the receiving cell site is constantly measuring the signal strength of the transmission. When the signal weakens, the transceiver in the cell site can send a power control message to the phone, instructing it to increase its power. If the signal increases, the same signal instructs the phone to decrease its power.

Another advantage of CDMA is the absence of crystals to support TDM. As we mentioned earlier in our discussion about

TDMA, crystals oscillate, creating potential RFI problems for other electronic equipment. Initial reports have shown little or no such interference from CDMA cellular telephones.

The handoff procedure used in CDMA is also different from other technologies. The TDMA air interface uses the standard hard handoff. CDMA uses the hard handoff, and it also has a soft one. When a cellular phone crosses cell boundaries, rather than using the hard handoff, the original cell continues to provide service to the phone. The new cell is also activated, and the phone operates over both cell sites until reaching enough signal strength that the new cell can take over. This is a complex technique, requiring both cell sites to be controlled by the MSC throughout the handoff.

The MSC is responsible for processing the transmission received from both cell sites and determining which cell site is in control at any one time. Although complex, the technique ensures reliable transfer of service during a call. There is no noticeable degradation in the transmission quality during the handoff (in TDMA and AMPS networks, static and volume drops are common during handoffs). This is especially critical for data transmission.

CDMA also provides support for data services. In today's CDMA networks, data transmission rates of 9.6 up to 14.4 kbps are supported. Future enhancements promise data rates of 64 and 500 kbps. Packet switching is also supported over the CDMA interface, without specialized packet-switching modems. Cellular digital packet data (CDPD) integration is supported as well.

Future enhancements to CDMA include video and high-speed data. This will be important to PCS network providers that want to offer enhanced features and services such as wireless videoconferencing.

Despite all of its features, CDMA hasn't been without its share of problems. Early deployment of CDMA experienced many problems when operating in areas with AMPS networks. An AMPS handset operates at 1 watt (W) and a fixed mobile station (installed in vehicles) can operate at up to 3 W. A CDMA telephone operates at 10 milliwatts (mW). AMPS units would interfere with and overpower the CDMA phones. This problem has since been rectified but has marred the reputation of CDMA nonetheless.

CDMA continues to be a debatable solution. Many companies are investing heavily in CDMA equipment, but there are just as many investing in TDMA. The next few years will determine who the winner will be. If CDMA is able to deliver what is promised, CDMA will clearly be the solution of choice for many wireless providers. If CDMA cannot deliver what it promises, the result will be financial disaster for those who invested in it.

12.3.3 Global System for Mobile Communications (GSM)

GSM development began in 1982 by the Conference Europeenne des Pasteet Telecommunications (CEPT). It was not until 1992 that the first GSM network was put into place in Germany. There were many influences in the development of GSM networks: the European economy, political pressure create an international network, and the absence of services to support Europe's changing workforce.

GSM did not start out as a digital solution. ISDN was already being deployed in Europe and had seen tremendous success. The Post, Telephone, and Telegraph (PTT) administrations of Europe had realized the many advantages of using ISDN over analog facilities and were very interested in expanding ISDN into their wireless networks. This was the principal reason behind making GSM a digital network.

ISDN is used for interconnecting MSCs, for both voice and data. Signaling in GSM networks uses another European technology, SS7. SS7 development had begun in the mid 1960s, and it was first deployed in Europe. Since GSM was all digital and SS7 was already serving the wireline network, it was a natural choice to use SS7 in the GSM networks as well.

ISDN and SS7 were the only existing technologies used as part of the infrastructure. All of the cellular-specific applications were developed specifically for GSM, including the air interface. TDMA is used as the air interface in GSM networks, which was adopted by the EIA/TIA for use in North America as well.

GSM provides a number of features now found in North American networks as well. Some other features of GSM are just now beginning to find their way into U.S. networks. Caller ID, call waiting, call hold, conference calling, and data rates of 2400 and 9600 bytes per second (bps) are supported by GSM.

Several types of cellular entities are used in GSM that are somewhat different from those in the United States. Terminal equipment (cellular telephones and modems) can be fixed, portable, or handheld.

A fixed mobile station is one that is permanently mounted into a vehicle. It can operate at up to 20 W, which is more powerful then its U.S. counterparts. The increased wattage provides more coverage but would introduce problems with interference if this were attempted in U.S. networks.

Portable stations are typically bag phones, which are larger than the units that fit in your pocket, yet small enough to still carry around. They are referred to as bag phones because they are installed in their own carrying satchels and operate at up to 8 W.

Handheld stations are much like the popular cellular phones used in the United States. They are very small and can be carried in a pocket. These stations operate at up to 2 W. Although they do not provide the power of portables or fixed stations, they are popular because of their very small size.

Another unique feature of GSM phones is the SIM. This is a credit card-like device that plugs into the GSM telephone and provides subscriber information to the network. Rather than program this information directly into the telephone (as is done in the United States), subscriber information is programmed into the SIM card, which can be used in any GSM phone. This feature allows GSM subscribers to rent phones and immediately activate them by simply inserting their personal SIM cards.

New applications for these SIMs are now being explored. European GSM providers are expected to deploy many new services, including services where credit is programmed into SIMs for airtime. When the airtime is used up, the card is no longer valid.

The components in the GSM network are similar to those found in U.S. networks. Security in GSM networks is a bit better than in U.S. cellular networks because of several added functions such as an authentication center, which is a database that checks the transmission for an encoded value to determine if the phone is valid or not. If the phone has been reported stolen, or if there are a number of suspicious transmissions (such as those covering a large geographical area), the transmission is either denied or flagged as suspicious for further monitoring.

The VLR also assigns a temporary phone number for the cellular phone when the subscriber changes networks (by roaming into a new area served by a different MSC) as a security feature.

GSM phones can also scramble transmissions, making it difficult to monitor GSM calls. This is an inherent security feature of GSM that many U.S. cellular providers favor. However, because GSM uses TDMA at the air interface, there are many limitations as well. It is still unclear whether or not GSM will enjoy widespread deployment in the United States in new PCS networks.

12.3.4   Personal communications services (PCS)

PCS is really not a new technology but a new feature set for cellular services. There are many dramatic changes introduced into the network to support PCS services. First of all, the frequency range is higher than what is presently allocated for cellular in the United States. PCS networks operate in the microwave band, which is 2 GHz. Many new service providers have jumped into the pool, hoping to reap the profits that cellular operators have enjoyed over the last decade. These new players face many obstacles before they can begin to realize these profits.

The FCC held several auctions in 1995 and 1996, allowing new service providers to bid for market areas, instead of simply allocating the frequencies for cellular usage. Many of these service providers have paid millions of dollars for their market areas, without even having a network in place. They now face millions of dollars worth of investments to build new digital networks to support the PCS services they hope to provide.

Making things even more precarious for PCS operators is the fact that cellular providers have already saturated these same market areas, offering free calls on weekends and free cellular phones (with extended contracts). To compete against the cellular network providers, PCS providers will need to first build their networks and then be able to offer the same coverage with extended features.

Cellular operators have been able to build their networks slowly since there was no other service like theirs. They spent 10 years building their coverage and negotiating access agreements with other cellular providers so that their customers could use their phones anywhere. PCS operators will not have the opportunity to spend this much time building their networks. PCS networks must be able to compete with cellular on the first day the network is turned up.

The technologies for PCS are all digital. TDMA, CDMA, and GSM are all likely candidates for PCS networks, as are IS-54, IS-136, and the other standards already used in cellular networks. The difference between PCS and cellular networks will be in the services offered. For example, paging services will allow you to use your PCS phone as a pager, with one unique difference. When you receive a page, you can either receive an alphanumeric message, or you can elect to connect to the party paging you. The caller will be placed on hold while you decide whether to answer the page by connecting or to ignore the caller and receive the alphanumeric page instead.

13

Synchronous Optical NETwork (SONET)

The telephone industry has been busy converting its existing outside copper plant to newer fiber optics. There are many advantages to using fiber rather than copper. Higher bandwidth is the foremost advantage; quieter communications is another. There are a number of other economical advantages to using fiber optics.

In the existing copper infrastructure, amplifiers are required at regular intervals. These amplifiers are placed in cabinets in various locations (outside large subdivisions, along highways). The amplifiers are analog devices, unless recently upgraded to support newer digital services (such as ISDN). The problem many telephone companies encounter when offering new services is in upgrading existing equipment (including these amplifiers and also multiplexers) to support the newer digital services.

With fiber optics, there are no amplifiers, only repeaters. The repeaters can be placed at greater distances from the central office and are easier to upgrade when necessary. If a fiber optic cable is cut, service can often be switched to another backup fiber. This of course depends on the network configuration. Most fiber optic operators are deploying SONET rings, which provide a self-healing network. Another advantage of SONET is its standard multiplexing format, capable of carrying DS1 transmissions or even DS3. SONET itself is a standard for fiber optics transmissions, making interconnection with other equipment easier. SONET also uses a flexible architecture, allowing for future growth.

One of the unique advantages of SONET is its ability to access a single transmission frame without demultiplexing the entire transmission. This is not always possible with the digital signal hierarchy. For example, if a company is using a DS3 transmission facility, and it needs to extract a single channel (DS0) from the DS3, the DS3 must first be demultiplexed down to DS1 levels. This is because special framing bits are inserted into the DS3 transmission, and they must first be extracted before the DS0 can be interpreted. The same is true if superframes are used. Special framing bits are inserted into the transmission stream, which must first be extracted before the DS0 can be accessed.

As seen in Fig. 13.1, SONET uses a more flexible architecture, allowing more diversity when interfacing to various networks. The illustration shows that any number of various signals can be input into a SONET device and can be extracted as easily. This means that a telephone company can easily incorporate SONET as a backbone network to support its many smaller networks.

SONET can also be used in smaller networks, such as in a campus environment. With the standardization of ATM, this becomes critical. ATM applications are bandwidth-intensive, requiring fiber optics to support them. For networks too small to warrant the cost of SONET, FDDI can be used in its place.

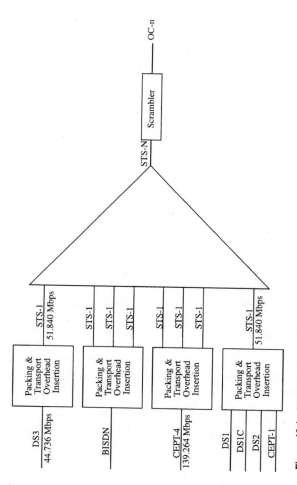

Figure 13.1 SONET multiplexing.

289

SONET can be deployed as either a bus topology (point-to-point) or in a ring configuration. Although there are many advantages to the ring topology, it is more difficult to manage from the protocol's perspective. This part of the standards is still improving as more and more companies deploy self-healing rings. The ring provides an additional level of reliability to any network, providing alternate routing in the event of failure. If a failure occurs, SONET nodes reverse the direction of traffic using the other ring.

13.1 SONET NETWORK NODES

Just a few devices are used in the SONET network compared to many different devices needed to support the digital signal hierarchy. In conventional digital transmission facilities, there are cross connects, channel banks, multiplexers, repeaters, and many other devices used in the outside plant. With SONET, the equipment list is reduced, representing reduced operating cost to network operators.

The add-drop multiplexer (ADM) can sit anywhere in the SONET network. It has the ability to drop any synchronous transport signal (STS) frame from within an optical carrier and route the STS frame onto another optical carrier or out to a non-SONET device. The multiplexer may sit in a central office where circuits need to be "plugged into" a SONET facility (which is where the "add" function fits). Figure 13.2 shows the basic function of an ADM.

A SONET terminal can receive signals from a variety of digital facilities and output a SONET optical carrier (see Fig. 13.3). It cannot pass traffic straight through or add or drop traffic from the optical carrier. The terminal is usually at either end

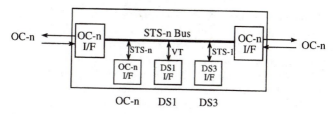

Figure 13.2 Add/drop multiplexer.

of a SONET point-to-point configuration, whereas the ADM is typically placed as an intermediate device (between terminals).

Digital cross connects are replacements for the old patch panels, where technicians used patch cords to manually connect one facility to another. The digital cross connect provides the same overall function, but there are no patch cords. A terminal is used to access software within the digital cross connect and configure which facilities are to be patched to which facilities. The digital cross connect is a real cost saver for all network operators because it simplifies and speeds the task of placing facilities into service and mapping them over the existing cable plant.

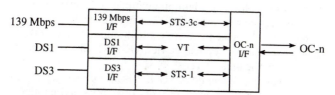

Figure 13.3 SONET terminal.

Synchronization in SONET is achieved through a hierarchical clock distribution network. The various levels in the clock distribution network are referred to as stratums, with stratum 1 being the most accurate signal available. There are five stratum levels. As signals are passed from one node to another within the exchange, the signal starts to degrade. As the signal becomes less accurate, it drops to a new stratum level. You could not use a lower stratum signal to feed a device requiring a higher stratum signal (for example, you could not use stratum 3 to drive a device requiring stratum 2).

Stratum 1 timing signals are usually generated by cesium clocks, which are traceable to Universal Coordinated Time (UTC). The GPS or LORAN-C is used to receive these reference signals, which are then distributed to clock cards for reference. The reference provides highly accurate timing phases (like the swing of a pendulum), to be used by the stratum 1 signal being generated. Figure 13.4 shows the clock hierarchy in relation to SONET nodes.

13.2   THE SONET PROTOCOL

The basic building block for a SONET frame is the 51.84-Mbps frame, as shown in Table 13.1. All of the rest of the SONET signal levels are built in multiples of this STS-1 level. In the table, you will notice two identifiers for each level. There is the STS side, which is the electrical input into the SONET multiplexer, and the optical carrier (OC), which is the optical output from the multiplexer.

Another unique feature of SONET is its ability to accept input from a variety of different types of sources and multiplex them into SONET frames. Figure 13.5 shows various types of

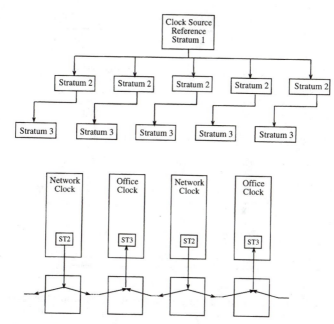

Figure 13.4   SONET clock synchronization distribution.

input, including varying levels of digital signals, all multiplexed into SONET frames at whatever bandwidth is available.

Yet another unique feature of SONET is the ability to extract a single transmission from the SONET facility anywhere in the network. In other words, if you are troubleshooting a SONET facility and suspect that a particular node may be causing the problem, you can attach a monitor to the facility and

TABLE 13.1   SONET/SDH hierarchy.

North American Designation			
Electrical Signal	Optical Signal	Data Rate (Mbps)	CCITT Designation
STS-1	OC-1	51.84	
STS-3	OC-3	155.52	STM-1
STS-9	OC-9	466.56	STM-3
STS-12	OC-12	622.08	STM-4
STS-18	OC-18	933.12	STM-6
STS-24	OC-24	1244.16	STM-8
STS-36	OC-36	1866.24	STM-12
STS-48	OC-48	2488.32	STM-16

read the transmission from just that node. This would be impossible to do in, say, a T-3 because the entire facility would have to be demultiplexed down to a T-1.

The SONET system is divided into four layers. Each layer is responsible for specific functions. The four layers are

- Photonic

- Section

- Line

- Path

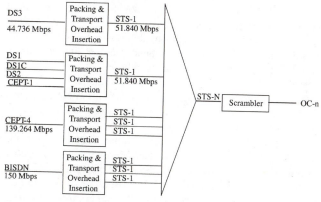

Figure 13.5 SONET inputs and outputs.

The photonic layer converts electrical signals to optical signals for output as an OC. This layer also performs the reverse, converting optical signals to electrical signals. The photonic layer is also responsible for managing the optical pulse shape, wavelength of the light, and power levels. SONET uses single-mode fiber but may vary in light strength and type depending on the distance to be covered.

The section layer defines the basic SONET framing to be used. It is responsible for communicating between adjacent signal regenerators as well as between regenerators and SONET terminals. It manages the transmission of multiplexed frames across the fiber, providing framing, scrambling, and error monitoring (of the section layer only). Section overhead is found in the first three rows of columns 1, 2, and 3.

The line layer provides synchronization and multiplexing of STS-1s between adjacent network nodes. Each node "terminates" the line layer, which is why nodes are referred to as line-terminating equipment. Maintenance at the line layer and protective switching is also managed here.

The path layer provides end-to-end transport of SONET frames. A logical connection is established between source and destination path-terminating equipment (a path is a logical connection between these two points).

The layers are somewhat similar to the various functions in a LAN. If you remember our discussions of LAN protocols, you will remember that there is the MAC layer, the LLC layer, and then the network layer (which is usually another protocol altogether). Routers use the network layer protocol information, while ignoring the data link protocol. Bridges, on the other hand, use the data link layer (MAC and LLC) while ignoring the network layer.

Figure 13.6 shows how the various layers in the SONET hierarchy communicate with various components in the network. The header for the path layer is read only by SONET terminals and is treated as part of the user data by regenerators and multiplexers. Likewise the line layer is only used by multiplexers and SONET terminals, and so on.

The various layers also have associated overhead. This overhead is appended to the user data and used by the various devices in the network, depending on the functional layer they represent. There is no overhead associated with the photonic layer. This is where the signal is output as light and is analogous to the physical layer of any other protocol. Whatever is passed through all of the layers above eventually makes its way to the lowest layer, whose job it is to transmit what it receives.

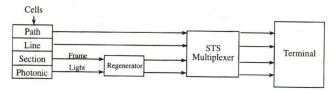

Figure 13.6   Logical hierarchy of SONET.

This is the photonic layer.

The section layer does come with overhead. The section overhead defines the SONET frame. The following parameters are part of the section overhead:

- *A1 and A2.*   Framing bytes used to synchronize the beginning of a frame.

- *C1.*   STS-1 identifier used to identify individual STS-1s within an STS-*n* frame. This is used when multiple STS-1s are interleaved into a larger STS-*n* frame (such as an STS-3). It is a binary number, assigned during multiplexing. The first STS-1 is assigned the value of 1 (0000 0001), and the next STS-1 is assigned the value of 2 (0000 0010).

- *B1.*   The parity byte used to provide even parity. It is used to check for transmission errors in a section. The value is calculated from all of the bits in the previous STS-*n* frame, after the previous frame was scrambled. The value is then placed in the first STS-1 of an STS-*n* frame before it is scrambled. This parameter is also referred to as the bit interleaved parity code (BIP-8) byte. Each parity byte is for error checking previous STS-*n*'s.

- *E1.* An optional 65-kbps voice channel used between "section terminating" equipment.

- *F1.* 64-kbps channel reserved for user purposes.

- *D1–D3.* These bytes are used as the section data communications channel (DCC). The 3 bytes provide a 192-kbps channel, which is used for alarm monitoring, control, and administration. A number of protocols are used to communicate over the DCC. The DCC protocol stack aligns with the International Organization for Standardization (ISO) OSI model. The next layer of overhead is the line overhead. The line layer points to the path overhead position in the frame. Pointers are used, rather than assigning fixed positions in the SONET frame, to allow for timing deviations that may shift the bit positions within the frame.

- *H1–H3.* Pointer bytes that are used in frame alignment and frequency adjustment of the payload.

- *B2.* The parity byte used for error monitoring at the line level. This is the same as the B1 parameter except that this is for checking transmission errors at the line level. The B2 parameter (and the BIP-8 code) is provided in all STS-1 frames within an STS-*n* (which is also different from the B1 parameter, which is only placed in the first STS-1 of an STS-*n*).

- *K1, K2.* Signaling bytes used between line-level switching equipment. A bit-oriented protocol is used on this channel to control automatic protection switching (APS). APS provides for an alternate path when there is a failure.

- *D4–D12.* A 576-kbps data channel for alarms, maintenance, control, monitoring, and administration at the line level.

- *Z1, Z2.* These bytes are used with BISDN applications at the UNI. They provide information on errored blocks of data (detected by BIP-8). Only Z2 of the third STS-1 within an STS-*n* is used.

- *E2.* A 64-kbps voice channel (PCM) for line-level order-wire. This provides a voice channel that can be used by maintenance personnel. It works somewhat like an inter-com, providing a 64-kbps voice channel over which techni-cians can communicate. The path overhead appears ahead of the data and indicates the demultiplexing format through the C2 byte. The path overhead parameters are

> *J1.* A 64-kbps channel; sends a 64-octet string at repeti-tive intervals. Receiving terminals use it to verify the in-tegrity of a path.
> *B3.* Parity byte used at the path level.
> *C2.* Signal label that is used to indicate whether or not a line connection is complete with a payload (equipped) or without one (unequipped).
> *G1.* Used by path terminating equipment to exchange status information.
> *F2.* A 64-kbps channel reserved for the user of the path level.
> *H4.* Indicates multiple frames were needed for the pay-load (segmentation indicator).
> *Z3–Z5.* Reserved for future use.

13.2.1 SONET framing

Now that we have covered all of the overhead presented in a SONET frame, look at Fig. 13.7 and see if you can identify all

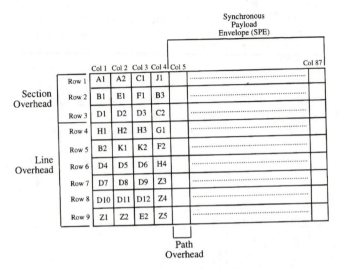

Figure 13.7 SONET frame.

the various layers of information. To present all of this information in one straight contiguous line would be rather difficult, so SONET frames are depicted in rows and columns. The actual transmission is a rather fast string of serial data.

The STS frame consists of 9 rows of 90 columns (each being 8 bits) for a total of 810 bytes. Byte transmission order is row by row, starting with the top row and moving from left to right. These frames are transmitted at the rate of 8000 frames per second, for a transmission rate of 51.840 Mbps.

The first three columns of each frame are overhead, and the remaining 87 columns are the synchronous payload envelope (SPE).

The SPE has its own format of 9 rows and 87 columns (transmitted row by row, left to right). The first column contains the path overhead, and the remaining 86 columns are for the payload.

A unique feature of SONET is its ability to allow the payload to "float" within an STS-1 frame. For example, if the payload is traveling a bit slower than the STS-1 itself, SONET allows for the slippage. This is accomplished by adding bits into the SPE portion of the STS-1 to compensate for the slippage. For example, if the payload slips in time, SONET will insert a byte at the beginning of the SPE. A pointer is then used to identify where the data within the SPE begins. This is adjusted at each node as the payload (user data) shifts within the SPE.

Both positive and negative bit stuffing are used to compensate for payload shifting. The SPE portion of the STS-1 is decoupled from the rest of the frame. This also allows the payload from any STS-1 to be dropped from any STS-n (or examined) for maintenance purposes. In negative bit stuffing, the payload is traveling faster than the STS-1. A byte of the actual payload data is inserted into the H3 field of the transport overhead, giving it a 1-byte advance.

13.2.2   Virtual tributaries

SONET was designed to carry 50 Mbps of data, but it was also designed to support existing digital facilities. A digital signal of any level could be mapped into a single STS-1, but this would be a waste of bandwidth. Virtual tributaries (VTs) allow multiple digital signals to be carried within one STS-1. There are different sizes of VTs (or types), each type supporting different digital signals (such as DS1, DS3, etc.).

The smallest VT is called VT1.5 and provides a data rate of 1.728 Mbps. This is used to support DS1s. A DS1 can be supported using a 27-byte structure. If you multiply the number of bytes by 8000 frames per second, multiplied by 8 bits per byte, you get 1.728 Mbps data rate ($27 * 8000 * 8 = 1.728$). There are four types of VTs:

- VT 1.5

- VT 2

- VT 3

- VT 6

Table 13.2 shows the capacity for each VT. They are often combined into groups to meet traffic demands. VTs are placed into groups to allow mixes of VTs within one STS-1. There can be seven different VT groups within an STS-1, each group using 12 columns of the SPE. A group cannot consist of mixed VT types. All of the VTs within a group must be of the same type. The capacity of a group depends on the type of VTs being supported. The following are the VT group types and their capacities:

- Four VT1.5s

- Three VT2s

- Two VT3s

- One VT6

Figure 13.8 shows how all of these VTs are sent into a multiplexer and then sent out as one STS-1.

TABLE 13.2   VT capabilities.

VT Type	VT Rate (Mbps)	# of Columns	Possible Payloads
VT 1.5	1.728	3	DS1
VT 2	2.304	4	CEPT-1
VT 3	3.456	6	DS1c
VT 6	6.912	12	DS2

Each of the inputs to the SONET unit is placed into a virtual tributary, which is then combined with other VTs and sent over the electrical interface as an STS signal. The multiplexer then combines all of the various STS signals and sends them to a scrambler, which converts the electrical signals into an optical signal.

Two modes can be used for VTs, locked and floating. If maximum efficiency of the 64-kbps DS0 structure must be maintained, locked mode is used. There is a one-for-one mapping, and the DS1s are not decoupled from the STS-1. This means that payload pointers are not needed and are available for payload instead. Locked mode of operation does not allow switching of VTs.

VT switching requires add/drop capability. A node is able to add or drop any VT within an SPE, switching it from STS-1 to STS-1. This function requires the VT to be in floating mode so that individual VTs can be accessed. This also requires the use of payload pointers so that the node can determine where the payload begins.

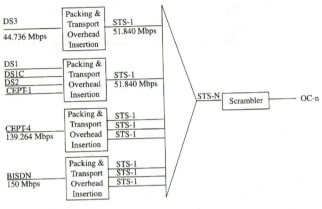

Figure 13.8   SONET signals.

13.2.3   Byte interleaving

Byte interleaving is the process used when sending multiple STS frames in an STS-*n* frame (for example, three STS-1s in one STS-3 frame). The bytes are interleaved into the STS-3 so that they can be removed from the STS-*n* individually. For example, in the case of an STS-3, the first byte of the first STS-1 is placed in the SPE of the STS-3. This is then followed by the first byte of the second STS-1, which is then followed by the first byte on the third STS-1. Three STS-1s can be interleaved into an STS-3.

The process is then repeated with the second byte of the first STS-1, and so on. The STS-3 can then be byte interleaved into an STS-12. Multiple STS-3s would be interleaved in the same fashion. The first byte of the first STS-3 is placed in the first byte of the STS-12, and so forth. It is byte interleaving that allows access to any portion of an STS-1 without demultiplexing the entire STS-*n*.

There are two modes of byte interleaving, single and two stage. With single-stage interleaving, there are no intermediate steps to larger STS-*n* frames. All STS-1s can be interleaved directly into an STS-12. Two-stage interleaving requires STS-1s to be interleaved into an STS-3 first, which is then interleaved into an STS-12. Figure 13.9 shows both methods of interleaving, single and two stage. The only requirement of single-stage interleaving is that the pattern must be the same as for two-stage interleaving.

In Fig. 13.9, the boxes represent the first byte of STS-1s. Notice the pattern when the first byte interleaved into an STS-12. Also notice that in two-stage interleaving, the same pattern is followed. This is so individual STS-1s can be found no matter what STS-*n* frame they are interleaved to. There are also bytes in the overhead used to correlate the STS-1s when they are received. The C1 byte of the section overhead is used to identify individual STS-1s when they are interleaved into a larger STS-*n* frame.

13.2.4 Automatic protection switching

An APS is part of the self-healing aspect of SONET. There are two types of APS, 1 + 1 and 1:*n* protection switching. With 1 + 1 protection, there are two paths for optical signals. The working path has a backup path, which is referred to as the protection facility.

The working facility is mirrored over the protection facility as if there were two different cables to the same termi-

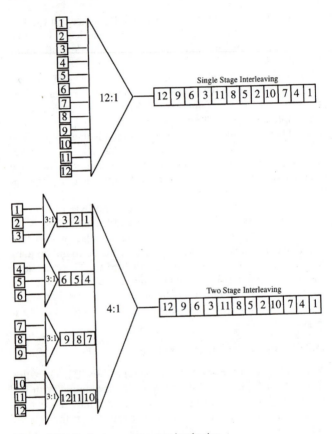

Figure 13.9   Single-stage and two-stage interleaving.

nation point. Everything that is transmitted over the working facility is duplicated over the protection facility. If a failure is detected at either end of the two facilities, the protection facility is immediately used. Since the same signal is already going over the protection facility, no "switchover" is required.

With 1:n protection switching, one protection facility is provided for multiple working facilities. Up to 14 working facilities can be supported by one protection facility. If any of the 14 working facilities fail, the signal from that working facility is placed over the protection facility. The protection facility can only support one working facility at a time, so if more than one working facility fails, there is a prioritization scheme for determining which working facility will be switched to the protection facility.

APS is used for point-to-point and point-to-multipoint configurations. With a ring topology, a different mechanism is used. The same bytes are used for ring protection switching, but the procedures carried out are different. The protection facility is the other ring in the dual ring configuration, and any node in the network can automatically switch traffic over to the backup ring.

This is just a quick overview of SONET and its capabilities. The future of telecommunications depends on the deployment of SONET networks, and many companies have been aggressively upgrading their networks to SONET networks to support the many services their customers are demanding.

There are a number of digital subscriber loop (DSL) technologies. Telephone companies are busy testing these now to determine which is the best solution for them. This is known as fiber to the curb (FTTC). An optical network unit (ONU) serves as the interface between the central office and the subscriber. The ONU provides optical-to-electrical conversion, voice conversion to PCM, and multiplexing functions.

The ONU serves up to 24 customers. The intent is to serve as few customers as possible, providing more bandwidth on the upstream channel for video conferencing and other interactive services. ONUs connect to a host digital terminal (HDT), located in the switching equipment in the central office. The HDT can also be deployed at remote sites. The function of the HDT is to control multiple ONUs and act as an interface to the rest of the PSTN.

By using FTTC and coaxial cable to the residence, telephone companies have positioned themselves to provide almost any service that subscribers need. Coaxial cable is capable of supporting high-speed data as well as video and voice, and with technologies such as ATM in the network, the telephone and the cable companies can support them all in one network.

There is also a cost benefit. Once deployed, there are no repeaters, network interface modules, or amplifiers to service. However, this also represents additional costs to telephone companies, which already have these elements in their outside plant. All of the outside plant must be changed to support FTTC, which raises the deployment cost significantly.

Another alternative to FTTC is hybrid fiber coax (HFC). This is a favorite among cable television operators because they already have a fiber optics backbone in place. However, HFC uses a bus topology, with drops to coaxial facilities that serves many users at once. The voice has to be multiplexed using frequency multiplexing because many conversations share the same bus. This represents a cost to the cable companies that want to use their existing networks for voice and data.

HFC is costly to deploy in the network because of the amount of outside plant that must be replaced. This is one reason why telephone companies have been slow to support HFC. They already have a substantial infrastructure to deal with, and it is too costly to try to convert it all. In contrast, cable companies are building their voice and data infrastructure from scratch, making it easier for them to invest heavily in fiber networks and coaxial drops to their subscribers.

The other advantage of HFC is that it supports both analog and digital signals. Cable companies can migrate to digital technologies much more slowly, while still providing the services they want to their subscribers. The next generation of HFC will support higher bandwidths but will require ATM switches as well as special digital set-top boxes.

The newer version of HFC will support hundreds of video channels as well as voice. With HFC, the fiber does not penetrate as deeply into the network as it does with FTTC. The fiber is run to distribution points, which may serve hundreds of customers. From the distribution point, coaxial drops are used to the subscriber. For this reason, HFC is sometimes referred to as fiber to the loop (FTTL).

Work is under way to bring fiber closer to the home using HFC technologies. The problem with the current implementation is that distribution points must support hundreds of subscribers. These subscribers must share the same upstream channel, which means video conferencing and other similar applications cannot be supported. If the fiber is brought closer to the subscriber, fewer subscribers will be served by a distribution point, providing more aggregate bandwidth per subscriber on the upstream channel.

Asymmetric digital subscriber loop (ADSL) uses the existing twisted pair at a residence, but it requires changes in the outside plant. A new line card supporting ADSL must be installed in the central office switching equipment, and all repeaters must be changed. Special set-top converters are required to convert video signals from digital back to analog for viewing.

ADSL splits the bandwidth of a copper loop into three channels:

- One downstream (to subscriber) 1.5- to 6.1-Mbps channel

- One upstream (to the central office) 16- to 640-kbps channel

- One voice channel

The downstream channel is capable of supporting near VHS quality video, using Motion Picture Experts Group (MPEG) compression. However, it is unlikely that broadcast-quality video will be supported for some time to come. If broadcast video is to be supported, the outside plant may prove inadequate due to its high noise potential.

A modem at the customer premise is used to multiplex these channels over the twisted pair. The modem is the line termination and is placed outside the home where the outside wire meets the house. If the service is more than 18,000 ft from the central office, a repeater is needed to support the signal. This adds additional cost to the service but is still cheaper than FTTC.

The modem uses discrete multitone (DMT) for modulation, a technique developed specifically for ADSL. This is only necessary if fiber is not used. The in-house wiring is sufficient if DMT is used, even allowing for a private LAN within each residence. In the event there is a power outage, POTS service is guaranteed by the ADSL modem. Figure 13.10 shows the reference model for ADSL, depicting the various functions required and the interfaces to those functions.

The ADSL transceiver unit/central office (ATU-C) is the line card used in the central office to provide the ADSL service to subscribers. This is used in the central office switch and does not imply a new piece of equipment to the telephone company (other than a new type of line card). The opposite end of this circuit is the ADSL transceiver unit/remote (ATU-R), which is the subscriber side of the line. The ATU-C converts analog signals received from within the switching network to an ADSL signal before transmitting to the remote end. The ATU-R provides the same functionality at the other end of the circuit.

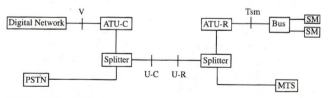

Figure 13.10   ADSI reference model.

The splitter can either combine or separate signals, depending on the direction of transmission. The digital network can be any kind of service. For example, connections to a video server would be through the V reference point to the digital network.

ADSL was first looked at for video-on-demand services. When the telephone companies discovered that there was no great demand for 200 channels, or video-on-demand, they quickly dropped the trials and began pushing ADSL as an Internet connection solution. Today, ADSL is looked at as an alternative to ISDN, supporting higher bandwidths at lower cost to telephone companies (typically under $1000 per home).

There are other versions of ADSL. One is symmetrical digital subscriber line (SDSL), which offers full duplex service. This means that there is a downstream channel and an upstream channel with equal bandwidth. SDSL provides T1/E1 speeds over twisted pair. High-bit-rate digital subscriber line (HDSL) provides higher bandwidths than ADSL but at a higher cost. Introduced in 1992, HDSL was used to deliver T-1 leased line services to businesses. The intent of the telephone companies was to use HDSL as a replacement for all of its T-span carrier equipment in the outside plant. This would lower their operating costs and provide more reliable service.

HDSL modems are cheaper to build than ADSL modems because they are not as complex. However, the overall deployment cost of HDSL is higher than ADSL. Although the bandwidth of HDSL is higher (6 Mbps downstream), some are still skeptical that HDSL will win over ADSL. There still are no requirements for that much bandwidth in the home.

The jury is still out on which of the fiber-in-the-loop solutions will be the winner. There is still time to weed out the weak entries and name the winner. But applications are changing, and as applications change, the ideal solution changes as well. It is hoped that the marketplace will find its calling, and the solution will be there.

14

Asynchronous Transfer Mode

ATM began development in 1969, when engineers at Bell Laboratories began experimenting with alternative switching techniques for routing data of all types through a common switching point. This would alleviate the need for multiple types of equipment in the central office and would provide a much faster, more efficient method by which data could be transmitted.

In the current telephone network, facilities are run from the central office to the subscriber over analog or digital lines. In most metropolitan areas, digital transmission is used up to the subscriber. This requires a multiplexer in the central office, as well as repeaters in the field. DS1s and DS3s are not used for facilities to the residential subscriber, so a different method of multiplexing is required.

These multiplexers (such as D4 and SLIC96) can be found in the field, close to the subscriber. However, using this method of distribution has proven costly to telephone companies due to maintenance and upgrades to support newer technologies. One example of this cost is ISDN, which cannot be used in many areas because the multiplexers used outside the central office will not support it. These must be replaced with compatible multiplexers before ISDN services can be offered. This has resulted in delays in providing ISDN in many rural areas.

ATM can provide a more efficient and cost-effective approach to distribution. By using one common facility throughout the network, telephone companies can rid themselves of the various pieces of equipment they own and maintain today. Using nothing but ATM switches and routers provides a standardized approach to transmission facilities.

ATM is capable of supporting any type of transmission, whether it is voice or video, making it an attractive choice for those companies that are looking to broaden their service offerings. Many telephone companies have already begun looking at cable television services and video on demand. Cable television companies, which already have a fiber optics network, have also been looking at ATM to deliver voice transmissions to their cable customers, competing directly with the local telephone companies.

ATM standards are referred to as BISDN in the ITU standards publications. BISDN is the next evolution of ISDN, providing higher bandwidth and faster speeds than ISDN. The ATM Forum has gone one step further and defined services for non-BISDN applications (data only). These applications are data-specific and include applications such as bridging LANs through ATM connections. BISDN standards define integration of data, video, and voice networks over one network. Integration is certainly not a new concept. The whole basis of ISDN was to create one common interface to support all forms of traffic, including data and video. ISDN fell short of delivering for a variety of reasons, and its acceptance has been slow because of a lack of killer applications.

ATM has already suffered some of the same maladies as ISDN. Slick marketing has touted ATM as the network solution, providing more bandwidth to the desktop than anyone could possibly need. Which was exactly the problem. Very few companies have expressed a need, let alone an interest, in having more than 100-Mbps bandwidth delivered to their desktops.

So why was ATM developed? The BOCs fostered the original idea long before divestiture. The idea of integrating their telephone networks generated the idea of ATM to replace the existing circuit-switched network. ISDN was the first step in integration, but as time evolved, the telephone companies recognized the need to increase the capacity of their backbone networks (as well as the local loop).

ATM was seen as the answer for backbone applications. By replacing the switched facilities with ones that made more efficient use of the bandwidth and supported all types of digital traffic, the telephone companies could consolidate their intraoffice facilities. ATM standards began with the ITU-T, in conjunction with the ANSI and now Bellcore. In 1991, four companies joined together to expedite the standards development process. Adaptive, Cisco, NTI, and Sprint formed the ATM Forum, with the intent of creating implementation agreements to be used as standards until the ITUs work was complete.

An implementation agreement is an agreement among all member companies to implement a technology in a specific way. It is much like a standard but is not officially drafted as one. The ATM Forum quickly grew to over 500 members. The ATM

Forum also streamlined the voting process in hopes of getting consensus on various contributions more quickly than the unanimous rule used by the ITU. The ATM Forum uses a majority rule, with a simple show of hands rather than formal balloting. What the ATM Forum did not count on were the problems in getting consensus among 500 different companies. Some companies have joined together and implemented their own solutions because they did not like the ATM Forum decision.

ATM is part of BISDN. BISDN is a combination of ATM, the ATM adaptation layer (AAL), and various protocols such as Q.2931 (signaling protocol derived from ISDNs Q.931). It uses a fixed-length cell of 53 bytes and is capable of carrying any type of data. Even data with protocol headers from other networks can be carried in an ATM cell (or segmented into several ATM cells).

The concept is that small cells introduce less delay. Devices waiting to transmit do not have to wait as long if cells are short. Small cells can be processed more quickly than large cells, reducing the processing at network nodes. The small cell size was a compromise between the telephone and cable industries and the data industry. Although large cell sizes are better for data, small cells work better for voice and video.

Various "planes" are supported in ATM. Think of each plane as a level of communications. The user plane transmits user data from one endpoint to another. This plane is also responsible for multiplexing among different connections (using the virtual path identifier/virtual channel identifier, or VPI/VCI) and for cell rate decoupling, cell discrimination, payload type discrimination, and traffic shaping. The user plane performs selective cell prioritization using the cell loss priority (CLP) field in the ATM cell header.

14.1   ATM PLANES

The control plane establishes virtual connections and handles signaling traffic. The management plane is divided into two parts: plane management and layer management. The management plane provides alarm surveillance, connectivity verification, and verification of VPI/VCI. Operations, administration, and management (OAM) cells are used to exchange alarm and connectivity information between nodes. If an invalid VPI/VCI is received, the cell is discarded and layer management is informed.

14.2   ATM SERVICES

Several ATM services support many of the applications we have already discussed. The services are

- Cell relay services

- Frame Relay bearer services

- Connectionless services

- LES

- ATM video and audio services

- ATM CES

Cell relay services provide point-to-point data transfer, without the services of the AAL. Cells are generated at the source, which means the source must be ATM equipment. It is believed that this will be the first usage for ATM networks, primarily supporting data transfers. However, it has become obvious that this may not be the case since many companies choose alternatives such as Fast Ethernet and FDDI. The ATM layer sends these cells over the network exactly as they were received and when they were received, making this the true ATM application.

Frame Relay bearer services requires the functions of AAL-5. This is an interworking solution for Frame Relay connections to the ATM network. The Class C functions of AAL-5 (and the various sublayers of the AAL) perform checks on the Frame Relay data to ensure that the length of the frame does not exceed what was negotiated at setup time and that the DLCI is active and assigned. This service may be used to interconnect two geographically separated Frame Relay networks. It may also be used to connect Frame Relay devices to ATM devices.

Connectionless services support the needs of protocols such as Switched Multimegabit Data Service (SMDS), Connectionless Broadband Data Services (CBDS), and Connectionless Network Services (CLNS), all examples of connectionless networks that would use ATM networks. They provide many features offered in LAN protocols on a WAN scope. They offer speed and less processing by network nodes (by eliminating error correction in the network).

The Data Exchange Interface (DXI) is part of this service. Defined by Bellcore, the DXI places the segmentation and reassembly (SAR) sublayer of the AAL in a separate device, such as a data service unit (DSU) or channel service unit (CSU). The purpose was to make ATM more cost effective for data applications. This is achieved by purchasing a software upgrade for routers, converting them to SMDS or CLNS. The router then sends PDUs in a special DXI frame to a DSU/CSU equipped with ATM AAL software. The PDU is then placed into an AAL frame, passed to the ATM layer in the DSU/CSU, and then placed into ATM cells for transmission over the network. The DSU/CSU has many input ports, so the cost can be shared by many departments instead of each department having to pay for its own ATM solution.

We have talked already about LES. It allows two different LANs to be connected as if they were one large LAN, even if they are geographically separated. This is an ideal solution for large corporations that want to maintain one large network without worrying about other protocols in the middle. Typically, a company may have Ethernet in its offices, Frame Relay interconnecting its offices, TCP/IP running on top of the Frame Relay providing e-mail over the Internet, and ISDN for voice applications. Using ATM it can reduce the number of interim protocols used for interconnectivity.

ATM video and audio services support the requirements of broadcast-quality video. We discussed some of the issues already when we talked about the video and audio applications. This service will meet the requirements of "raw" video, compressed video, and audio tracks that accompany the video.

14.2.1   ATM bearer services and classes of service

ATM provides a connection-oriented cell transfer service between a source and a destination (end to end). Sequenced delivery is guaranteed. ATM also supports quality of service and throughput by negotiating these parameters at connection establishment. Switched virtual connections (SVCs) rely on the services of the control plane, adaptation layer, and signaling protocol. The four classes of service defined are

1. Class A-Circuit emulation, constant bit rate (CBR) video
2. Class B—Variable bit rate (VBR) audio and video
3. Class C—Connection-oriented data transfer
4. Class D—Connectionless data transfer

These class types provide a means for delivering the services defined. This is different from the AAL types, which provide a service. The AAL identifies the type of class to be supported and other services needed to deliver the data.

14.3   ATM NETWORK ACCESS

Two interfaces are defined in ATM standards, UNI and NNI. The UNI is the best defined and has been the focus of the ATM Forum for the last few years. This is the interface between the public network and the private network (or ATM user). The NNI is still evolving and is the interface between public networks. The ATM header is different between these two interfaces, and the services provided differ as well. Figure 14.1 shows the various ATM interfaces and their relationships.

Both the UNI and the NNI are defined as public or private. Public interfaces are provided by service providers and require more complexity than private interfaces. A private interface exists within a closed network, such as within a large corporation. Although at some point access to a public interface may exist, the private interfaces link only to internal nodes (within the private network).

14.3.1   User-to-network interface

As mentioned above, UNI is the interface between the ATM user and the ATM public network. Because ATM is expected to support a variety of different types of applications, it should

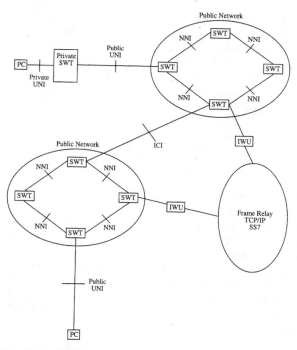

Figure 14.1   ATM interfaces.

also be assumed that ATM must support a variety of different network types at the subscriber premise. The AAL was developed to interwork different network types with the ATM network and maintain transparency to the user.

The beauty of this concept is that subscribers can continue using the LAN technologies they already have in place, while enjoying the benefits of ATM access to the public network. All of their telecommunications networks, data, voice, and video, can be merged at the public network and transferred to the public network over a single ATM facility. The alternative is to lease different types of facilities to support each unique application. This would mean using a channelized service such as T-1 for voice applications, possibly a leased line or Frame Relay service for data, and a leased line or ISDN for the video network. The cost of this approach is somewhat prohibitive, not only for the subscriber but also for the service provider, which must maintain several pairs of copper to the subscriber premise to support all of these services.

ATM is advantageous then to both the subscriber and the service provider, providing facility consolidation as well as additional bandwidth to support a variety of broadband applications. The key here is reduced cost and enhanced services as much as it is more bandwidth.

The physical interface for the UNI is currently defined as 44.736-, 100-, and two different 155.52-Mbps interfaces. The 155.52-Mbps interface uses SONET STS-3c. The NNI interfaces range from 155.52 up to 600 Mbps (in concert with SONET).

14.3.2 Network-to-network interface

The NNI is used within the public network. This is where the various service providers interwork their networks with one another. NNI provides a different set of services than the UNI does simply because the requirements are very different. The UNI does not require the same types of network management and reliability that the public network does.

The adaptation layer is not supported over the NNI, mostly because it is not needed. The NNI is simply interconnecting ATM networks to one another and does not require interworking with unlike networks.

Applications within the public network, such as SS7, use the adaptation layer. The SS7 network must be able to connect to the ATM network but will remain a separate network for now. Work is under way to develop interfaces for SS7 signaling points, supporting CES and AAL for the transport of signaling messages over the ATM network. There are many cost advantages to this, but the principal reason is twofold. As the ATM network migrates into the public network, replacing many of the interoffice facilities, the circuits used for SS7 will also be replaced. At some point, SS7 links will need to be placed on ATM facilities because the channelized facilities will be gone. The SS7 signaling points will remain an integral part of the network, especially the STP and SCP because of the services they provide to the public telephone network.

Another factor is the amount of traffic in the SS7 network. In today's network, SS7 is able to support the capacity. However, new applications are being deployed for SS7, and the traffic mix is changing rapidly. More and more database applications require SS7 services, which means SS7 traffic is increasing. This increase will soon begin taxing the 64-kbps links that are used today. ATM allows the telephone companies to integrate their interoffice facilities to carry all traffic, including SS7 signaling traffic.

14.4 OAM MESSAGES

The OAM cell uses a different format than other cells. Two types of cells are supported, F4 and F5. Figure 14.2 shows both

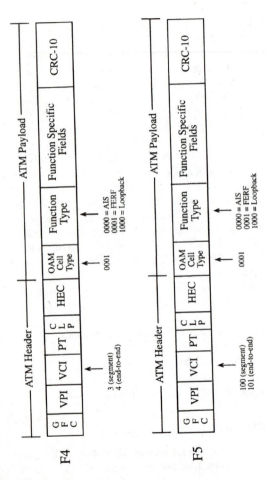

Figure 14.2 F4 and F5 OAM cell formats.

formats. Two types of indications are given through OAM alarming: alarm indication signal (AIS) and far end receive failure (FERF). Both types provide a failure type and failure location in the "function-specific" field of the OAM cell.

The AIS message is used to alert downstream nodes of an alarm condition. Either virtual path connection/virtual channel connection (VPC/VCC) failure or physical layer failure can cause AIS message generation. When a switch at the public UNI receives an AIS message, it returns an FERF message to alert the downstream nodes of the failure downstream. A virtual path AIS and virtual path FERF are always carried on VCI54. A virtual channel AIS and virtual channel FERF are always carried over cells with PT5101.

14.4.1 OAM connectivity verification

OAM payload consists of loopback indication, a correlation identifier, loopback location identifier, and source identifier. Loopback indication indicates that the cell should either be looped back or discarded. If looped back, the value is decremented (to prevent the originator from interpreting it as a loopback from another node and sending it back). Correlation ID allows the originating node to keep track of OAM responses when OAM cells are sent over the same virtual connection. This value is used only by the originator of an OAM cell and is ignored by other nodes.

Loopback location ID is a 96-bit field identifying where the loopback should occur within a virtual connection. A value of all 1s indicates that the loopback should occur at the remote endpoint. Source ID allows the originator to identify the loopback instruction as one it has sent. No values are defined, and the standard allows any kind of identification.

Segment loopbacks take place between the subscriber equipment and public ATM switch on either end of the UNI. In other words, the segment is defined as the link between these two points. The loopback message never travels beyond these points.

End-to-end loopbacks can be generated by any node in the connection but can only be discarded or looped back by endpoints (defined by the VPC or VCC). These messages are identified as

- Payload type 5101 for VCC (end to end)

- VCI54 for VPC (end to end)

- Payload type 5100 for VCC (segment)

- VCI53 for VPC (segment)

Loopbacks are used to ensure that a logical connection can be established. In analog facilities, a connection must be established over the facility before any connection requests can be made with the remote end. This is when analog facilities are tested for continuity. In digital facilities, there is no way of testing a connection prior to connection establishment because there is no physical connection, only virtual connections. Signaling is often sent over different channels. Loopbacks over digital facilities are sometimes called *continuity checks*.

14.5   INTERIM LOCAL MANAGEMENT INTERFACE

Interim Local Management Interface (ILMI) provides status, configuration, and control information about links and physical layer parameters at the UNI. It also provides address registration across the UNI. ILMI is a protocol used between two adjacent UNI management entities supporting bidirectional communications between the nodes.

Each ATM device is associated with a UNI management entity that resides within software in each device. The management entity interacts with software which resides in network management stations. These stations typically exist in locations that are remote from the ATM nodes and are capable of communicating with the ATM switches throughout the network.

14.6   ATM LAYERS

ATM is divided into layers (see Fig. 14.3). The physical layer is divided into two parts. The ATM physical medium sublayer is responsible for the transmission of data over the physical medium, regardless of the type of medium used. ATM was originally designed to operate over fiber optics but because of the slow deployment of fiber, was later modified to operate over copper and coaxial facilities as well.

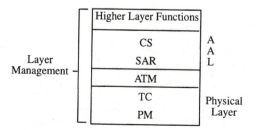

Figure 14.3   ATM layers.

The physical medium sublayer is responsible for receiving and transmitting bit streams in a continuous method. This is important to channelized services, which rely on constant bit streams to maintain synchronization. When the bit stream stops, channelized equipment interprets the condition as an error and releases the virtual connection. This sublayer also maintains bit synchronization.

The transmission convergence sublayer is responsible for the transmission and reception of frames over a framed transmission facility, such as T-3. ATM cells are packed into these frames and unpacked at the remote end. This sublayer also performs error detection/correction but only on the ATM header. This prevents the cells from being sent to the wrong destination.

Cell rate decoupling is used when a continuous data stream is required at the physical layer, as in SONET and channelized facilities such as DS1. Cell rate decoupling sends special "idle" cells over the framed facility and discards any idle cells it receives. Idle cells are necessary to maintain a connection in channelized facilities because the channel bank equipment must always see a constant bit rate transmission, or it disconnects the channel. When nothing is being sent over a channelized facility, idle flags are transmitted (this is also used to maintain clock synchronization between two endpoints). The ATM layer does not recognize idle cells.

The functions of the transmission CS differ depending on the medium being used. For instance, if SONET is the medium, the physical layer requires a different set of functions than a DS3 medium would require. This sublayer provides whatever services are needed by each type of medium.

Some specific functions are required for DS3 and 100-Mbps interfaces. The ATM physical layer provides a convergence protocol (Physical Layer Convergence Protocol, or PLCP), which maps ATM cells onto a DS3. The interface supports 44.736 Mbps. ATM cells are mapped into a DS3 PLCP data unit, which is then mapped into the DS3 payload. The DS3 PLCP is not aligned to the DS3 framing bits.

The 100-Mbps access was intended for private UNIs, which are not as complex as public UNIs, which must provide higher reliability and complex monitoring. The specification is based on the FDDI physical layer.

It was mentioned earlier that two 155-Mbps interfaces were defined. One is for the public UNI, and the other is for the private UNI. The difference lies in the distances supported by each interface. The 155-Mbps private UNI interface can be used over fiber optics or twisted pair copper. The public UNI requires fiber optics using single-mode fiber.

The ATM layer is responsible for multiplexing cells over the interface. ATM must read the VPI/VCI of incoming cells, determine which link cells are to be transmitted over, and place new VPI/VCI values into the header. At endpoints, the ATM layer generates and interprets cell headers (endpoints do not route cells).

The ATM layer supports the following connection types:

- Point-to-point virtual channel connection (VCC)

- Point-to-multipoint VCC

- Point-to-point virtual path connection (VPC)

- Point-to-multipoint VPC

A VCC is a single connection between two endpoints. A VPC is a bundle (or group) of VCCs carried transparently between two endpoints.

The AAL is used mostly by endpoints. It is divided into two sublayers: SAR and the CS. The SAR reconstructs data that has been segmented into different cells (reassembly). It is also responsible for segmenting data that cannot fit within a 48-byte payload of an ATM cell (segmentation).

The CS determines the class of service to be used for a transmission. This will depend on the bit rate (constant or variable bit rate), the type of data, and the type of service to be provided (connection-oriented or connectionless). The quality of service parameters necessary for the transmission are determined by the class of service assigned.

14.6.1 ATM header and payload

The ATM cell is 53 bytes long. The first 5 bytes are used for the header. The payload portion of the cell is 48 bytes. Keep in mind as we discuss the protocol that the payload is also used for the AAL overhead and any other overhead from upper layers.

There are two formats for the ATM header. One header is used for the UNI, and the other is used for the NNI. The difference lies in the generic flow control (GFC) parameter found in the UNI header. The GFC is not supported in the public network, nor was it intended to be.

The header is placed in front of the payload (it arrives first). No trailer is used in ATM. Figure 14.4 shows the two-header formats. In the UNI header, the GFC can be used to throttle traffic from a specific destination. The GFC values are not currently defined, but the intent is that the GFC could be used to provide flow control from the user to the network (and vice versa). This parameter is not used out in the public network and is overwritten by network nodes.

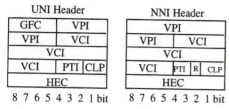

Figure 14.4 ATM headers (UNI and NNI).

Two modes are defined for GFC, controlled and uncontrolled. Controlled GFC allows subscriber equipment to control the flow of ATM cells; however, the values for this have not been defined. They are of local significance only, which means they are related to a link connection at a switch and used to communicate with the remote end of a link. Uncontrolled GFC simply means that GFC is not supported, and the parameter is set to all zeroes.

The VPI is used to identify a group of virtual channels with the same endpoint. This is the form of addressing supported in ATM. Rather than identifying millions of unique nodes, ATM addressing identifies a connection. A virtual channel is used for a communication link. Each virtual channel is identified by the VCI.

The meta signaling channel is a dedicated channel used to establish virtual channel connections between two endpoints. The virtual paths are predetermined at the time of installation (either through administration or by the service provider). Virtual paths can be negotiated by the user or the network using meta signaling. This means the establishment or release of virtual paths can be controlled by the user or the network.

Following the VPI and the VCI is the payload type indicator (PTI). This parameter indicates the type of data found in the payload portion of the cell. Remember that the payload is not always data; it could be signaling information, network management messages, and other forms of data. These are identified by the PTI.

The PTI is followed by the CLP parameter. This is used to prioritize cells. In the event of congestion or some other trouble, a node can discard cells that have a CLP value of 1 (considered low priority). If the CLP value is 0, the cell has a high priority and should only be discarded if it cannot be delivered.

The last byte in the header is the header error control (HEC) parameter, which is used for error checking and cell delineation. Only the header is checked for errors. The HEC works like other error-checking methods, where an algorithm is run on the header and the value placed in the HEC. ATM is capable of fixing single bit errors but not multiple bit errors.

An ATM node places itself in error-correction mode during normal operation. If a single bit error is detected in the header, the data in the header is small enough that the error-correction algorithm can determine which bit is in error and correct it. If a multibit error is detected, the node places itself in error-detection mode. Errors are not corrected while in this mode. The node remains in error-detection mode as long as cells are received in error. When cells are received without error, the node places itself back into error-correction mode.

14.6.2   Routing in ATM—VCI/VPI

In circuit-switched facilities, time slots are assigned to a transmission (or to a device). When a device wishes to transmit data, it must wait for its assigned time slot. The assignment of time slots can be dynamic, as is the case for ISDN. This means that a multiplexer assigns a time slot when a device requests a connection over the network.

In ATM, a specified number of cells are made available during a time period. A device can take any available cell to transmit data (or multiple cells, as is typically the case). The ATM multiplexer takes the data received from the device, adds the 5-byte header, and transmits a cell. In some cases, the data may first have to be processed by the AAL, which then adds a header and trailer and passes the frame to the SAR, which then segments the data into several data units. The data units are then sent to ATM, where the header is added.

ATM addressing consists of two identifiers, which identify the virtual path and the virtual connection. Figure 14.5 shows the relationship between virtual channels and virtual paths. A virtual path consists of multiple virtual channels to the same endpoint. The virtual channels are static to a destination. The virtual paths change at each node and are prearranged either through system administration or by a signaling protocol. This is analogous to a trunk group assignment in voice switches, where the trunk group consists of multiple voice circuits going to the same endpoint, each with its own unique identifier. A connection from end-user VCI to end-user VCI is called a VCC. A connection from end VPI to end VPI is called a VPC.

The UNI supports a maximum of 256 VPIs at any one node for each physical link. Remember that each VPI is an ATM link connection, with multiple virtual channels. A maximum of 64,000 virtual channels is supported for each virtual path. The NNI supports a maximum of 4096 VPIs because the NNI ATM header does not have the GFC parameter. Instead, these 4 bits are used to expand the VPI.

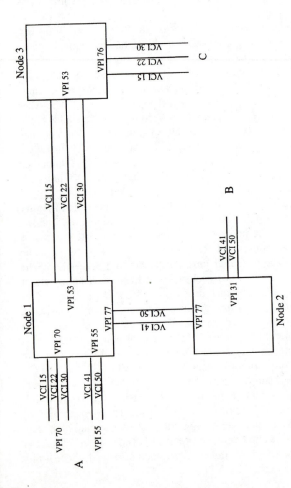

Figure 14.5 Virtual connections in ATM.

14.7 ATM SIGNALING

ATM meta-signaling is used for dynamic connections. These are made on demand and are released when transmission is complete. Both point-to-point and point-to-multipoint configurations can be supported. The alternative to dynamic connections is permanent connections, which are established at the time of installation. A permanent connection is just that; it remains connected all of the time, unless there is a failure. This is analogous to permanent virtual circuits.

Q.2931 defines the signaling messages used to establish, maintain, and release connections at the UNI. The public network will not use Q.2931; it uses SS7 instead. The SS7 protocol supports Q.2931 through the BISUP. Q.2931 was derived from the ISDN signaling protocol, Q.931.

ATM requires much more complexity than existing signaling. For example, if a caller initiates a voice call, a signaling message is generated to establish a connection for the voice. If the caller then activates a camera for video conferencing, another connection must be established for just the video segment of the call. Both connections must be correlated and synchronized.

Broadcast signaling virtual channels support connection establishment for applications where the same data must be sent to multiple destinations. There are two types, general and selective. General allows signaling to be sent to all endpoints at the user interface (not all endpoints within the network). Selective allows the network to broadcast signaling to endpoints that meet a particular service profile. Both general and selective are unidirectional, sent from the network to endpoints at a user interface. This means that ATM could place significant demand on the existing SS7 network used within the NNI. The BISUP protocol has been defined to support these ATM services and is still evolving. This is also the reason for expanding the capacity of the SS7 links beyond the present 64 kbps (in addition to requirements being placed on SS7 to support more database applications).

Many ATM books declare that SS7 is not required anymore or that it is unclear what role SS7 will play in the ATM world. However, the RBOCs are already busy planning expansions to their SS7 networks because they envision increased demand on SS7 services. The SS7 network provides much more than just connection establishment through the public network. It also supports database access for telephone switches providing intelligent network solutions, cellular applications, and now local number portability.

SS7 may remain a mystery to many in the data world mainly because it has always been an obscure telephone company technology. It is best understood by those directly involved with telephone networks (and in many cases not well understood even there). One thing is certain, SS7 is not going away, and ATM signaling will not replace SS7. ATM signaling is designed to meet the signaling requirements at the UNI, but it does not provide the services required of the NNI.

14.7.1 ATM addressing

Addressing endpoints in an ATM network is part of the UNI signaling. Although ATM switches route based on connections (identified by the VPI/VCI), once the cell arrives at the destination network, there has to be an address to get it to the proper node.

Endpoint addresses are carried in the payload of an ATM cell as part of the signaling message. Three different address formats are presently defined for use at the private UNI and an additional format for the public UNI:

1. DCC (private)
2. International code designator (ICD) (private)
3. E.164 ATM private format (private)
4. E.164 ATM public format (public)

Figure 14.6 shows the four address formats used in ATM. The authority and format identifier (AFI) defines the authority responsible for address registration and the format used. The authority can be an ATM equipment manufacturer, service

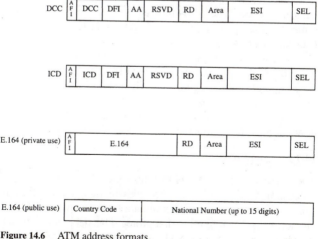

Figure 14.6 ATM address formats.

provider, telephone company, or administrator of a private network. The DCC is a 2-byte field that identifies the country in which the address is registered.

The domain specific part identifier (DFI) is a 1-byte field that specifies the structure of the rest of the fields. The administrative authority (AA) is a 3-byte field that identifies the authority that is responsible for the rest of the address. The routing domain (RD) field is a 2-byte field that specifies a unique routing domain. The area field is a 2-byte field used to identify an area within a routing domain.

The end system identifier (ESI) is a 6-byte field that identifies an end system within an area. The selector field (SEL) is a 1-byte field used by the end system to select an endpoint within an end system. The ICD is a 2-byte field that identifies an international organization. The British Standards Institute maintains these codes. The E.164 field is an 8-byte field that uses the same addressing as defined for ISDN; it is used to identify ISDN numbers.

When a cell is transmitted, the subscriber equipment provides the ESI and SEL values, which identify the end system and endpoint originating the cell. The network then fills in the rest of the address information when the cell is passed to the network over the UNI. The addresses are registered with the network for future connections.

14.7.2   Adaptation layer

The purpose of the AAL is to interwork ATM networks with other non-ATM networks. It provides the necessary information to ensure proper handling of data from a variety of different networks. Keep in mind that protocol headers remain intact in some cases, treated by AAL as part of the user data. For example, bridging an Ethernet LAN with another geographically separated Ethernet LAN requires the header from the Ethernet packet to be passed through the ATM network to the remote Ethernet network for proper routing in that network.

The AAL resides above the ATM layer. No cells are formed at this layer, only PDUs that are then passed to the ATM layer to be inserted into ATM cells. AAL is capable of viewing user data and is responsible for the segmentation and reassembly of user data.

There are two parts to the AAL. The convergence sublayer is the upper portion of the AAL. User data is first passed to the convergence sublayer, where it is encapsulated into a convergence PDU and passed to the lower part of AAL. The lower part of AAL is the SAR sublayer. The SAR is responsible for segmenting the user data, as well as any overhead added by the convergence sublayer, and for passing the segments to the ATM layer for encapsulation into ATM cells.

The convergence sublayer is divided into two parts (see Fig. 14.7). There is the service-specific convergence sublayer (SSCS) and the common part convergence sublayer (CPCS). User data is first presented to SSCS, where it is encapsulated into an SSCS PDU. SSCS provides clock recovery and message identification. The CPCS provides message identification and detection of sequence errors. These are still evolving standards.

Figure 14.8 shows the services provided by the AAL. A class is a means of providing a service, whereas the AAL type defines the service to be provided. Layer 2 of the AAL provides CRC and length identification of user data, buffer allocation size, and sequence numbering. Layer 3 provides buffer allocation size and sequence numbering. Layer 4 provides sequence numbering, segmentation, and reassembly.

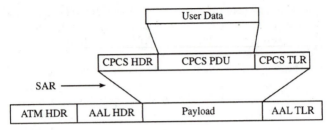

Figure 14.7   CPCS and SAR.

Additional overhead is associated with the AAL; it depletes the amount of available payload in the ATM cell because the overhead becomes part of the payload itself. In other words, AAL overhead has nothing to do with the ATM header. It is added to the user data in the form of a PDU, passed to the SAR sublayer for segmentation, and then passed down to the ATM layer for transmission as a series of cells.

AAL-1 is used for synchronous bit streams. Designed for voice and data transmitted over channelized facilities, such as T-1, AAL-1 delivers timing information from the source to the destination. AAL-1 provides segmentation and reassembly, management of cell delay variation, management of lost and misinserted cells, clock frequency recovery at the destination, and bit error recovery.

AAL-2 has been the most difficult to develop. The challenge has been defining a way by which AAL-2 can recover clock frequency at the destination when there has been a long idle period at the source. In synchronous networks, constant transmission of bits is necessary to maintain clock synchronization. When there are extended idle periods, clock synchronization must be reestablished.

	Class A	Class B	Class C	Class Y	N/A	Class X	Class D
	CBR			VBR			Connectionless
			Connection-Oriented				
	Timing Preserved		Variable Delay Acceptable				
Higher Layers	Any	Any	Frame Relay TCP/IP	Any	Q.2931	N/A	SIP-3, Others
Use	Circuit Emulation	VBR Voice, Video	Connection Oriented Data	Available Bit Rate	Signaling	Cell Relay	Connectionless Data
	AAL 1	AAL 2		AAL 5		AAL 0	AAL 3/4
Payload	47 bytes	45-47 bytes		48 bytes			44 bytes

Figure 14.8 AAL services.

This is not an issue with AAL-1 because there is a constant stream of bits. In AAL-2, video compression at the source eliminates a lot of the bit stream, resulting in a bursty-type traffic pattern. MPEG-2 (a video compression scheme developed by the ISO Motion Picture Experts Group) has a 10:1 compression rate (worst case). If there are no changes in the video scene, MPEG-2 can deliver ratios as high as 50:1, which means long idle periods. AAL-2 is still under development with both the ITU-T and the ATM Forum.

AAL-3/4 was at one time two separate services. They were combined because the only difference was the presence of one field in the protocol. AAL-4 used a 10-bit field to deliver connection-oriented services, and AAL-3 did not use this field. This was because developers created two distinct sets of code for both services, when one set of code could perform both functions.

AAL-3/4 provides connection-oriented and connectionless services for data transfer. This service was intended for applications that are not sensitive to delay variations. It is thought that the majority of applications using ATM will be data applications (especially in the early implementation of ATM). This may not be the case, as the focus of ATM shifts further away from desktop or LAN deployment to backbone applications such as the telephone network.

AAL-5 is sometimes referred to as the simple and efficient adaptation layer (SEAL). This layer does not provide sequencing or error checking (of user data). Instead, it relies on the upper layers to provide these services. AAL-5 supports connection-oriented, variable bit rate, timing-insensitive data. It is intended for use in point-to-point ATM configurations.

ATM is not a perfect solution for data, nor is it a perfect solution for voice. There is still a lot of work to be completed before the standards are finished. Once standards work has been stabilized, and vendors begin selling their wares in the main marketplace (rather than in trial networks), the real work begins.

Telephone companies run painstaking interoperability tests on all equipment they place in their networks. No equipment is placed in the public network and simply turned on. Tests must prove that the equipment will not introduce any errors into the network and will not cause any outages in it. All of this testing takes time, and solving problems discovered during testing also takes time.

What all of this amounts to is a long time between protocol development and deployment. It may be many years before ATM replaces the existing public network infrastructure and becomes as common as channelized facilities are today. Vendors are already busy developing their products, testing them in trial networks, and laying the groundwork for nationwide and international deployment.

APPENDIX

History of Computing

1614	John Napier develops algorithms.
1617	John Napier introduces "Napier's Bones."
1642	The Pascaline is invented by Blaise Pascal.
1804	Jacquard's loom is invented by Joseph Marie Jacquard.
1822	The difference engine is invented by Charles Babbage.
1834	Analytical engine is invented by Charles Babbage.
1847	Boolean algebra is introduced by George Boole.
1853	A working difference engine is built by George Scheutz.
1890	The punch card tabulator is invented by Herman Hollerith.
1906	The vacuum tube is invented by Lee DeForest.
1917	Frequency division multiplexing (FDM) is invented by Bell Laboratories.
1928	Eighty-column punch cards are introduced by IBM.

1930	The differential analyzer is invented by Vannevar Bush of MIT.
1940	The complex number calculator is invented by George Stibitz of Bell Labs.
1942	The differential analyzer is modified.
1942	The first electronic digital computer is invented by John Atansoff and Clifford Berry.
1943	The Colossus computer is invented.
1944	Harvard Mark I is invented by Howard Aiken and IBM.
1946	The ENIAC is invented by John W. Mauchly and J. Presper Eckert.
1947	The transistor is invented by William Shockley, Walter Bratlain, and John Bardeen of Bell Labs.
1948	The selective sequence electronic calculator (SSEC) is invented by IBM.
1948	The term *cybernetics* is first coined by Norbert Weiner.
1949	Magnetic tape storage and the BINAC are invented by Eckert and Mauchly.
1949	The Manchester Mark I is invented.
1950	The ERA 101 is invented by Engineering Research Associates.
1951	Lyon's Electronic Office (LEO) is invented by Lyon's Moving and Storage.
1951	The UNIVAC is introduced by Sperry Rand.
1951	The junction transistor is invented by William Shockley.
1952	The integrated circuit is invented by G. W. A. Dummer.
1953	Magnetic core memory is invented.
1954	Silicon-based junction transistor is invented by Texas Instruments.

1956   The first operating system is invented by Bob Patrick (General Motors) and Owen Mock (North American Aviation).

1957   FORTRAN is invented by John Backus of IBM.

1958   The integrated circuit is improved and introduced by Jack St. Clair Kilby of Texas Instruments.

1958   The 7000 Series Mainframe is introduced by IBM.

1960   The CDC 1604 is introduced by Seymour Cray of Control Data Corporation.

1960   The dataphone modem is introduced by Bell System.

1961   The Unimate is introduced by Unimation.

1962   Metal oxide semiconductor (MOS) transistors are invented by RCA.

1963   The ASCII standard is published by ANSI.

1964   The System 360 is introduced by IBM.

1964   BASIC is introduced by Thomas Kurtz and John Kemeny.

1965   The PDP-8 is introduced by DEC.

1967   The first RAM chip is introduced by Fairchild Semiconductor.

1968   CMOS is introduced.

1968   The mouse is invented by Douglas Engelbart.

1969   The RS232 standard is published by Bell Labs and the EIA.

1969   UNIX is introduced by Kenneth Thompson and Dennis Ritchie of Bell Labs.

1969   The ARPANET is deployed by the Department of Defense.

1970   The first ATM machine is installed.

1970   Light-sensitive chips (CCD) are invented by Willard Boyle and George Smith of Bell Labs.

1970	PASCAL is introduced by Niklaus Wirth.
1971	The microprocessor is introduced by Ted Hoff of Intel.
1971	The 8-inch (in) floppy disk drive is invented by Alan Shugart and IBM.
1972	The HP35 Electronic Slide Rule is introduced by Hewlett Packard.
1972	The first PONG game is invented by Nolan Bushnell of Atari.
1972	The C programming language is introduced by Dennis Ritchie of Bell Labs.
1972	The Intel 8008 is introduced by Intel.
1973	The Winchester Disk Drive is introduced.
1973	Universal Product Code (UPC) is introduced.
1973	The Ethernet LAN is introduced by Robert Metcalfe of Xerox.
1974	The Intel 8080 CPU is introduced.
1974	The Z80 CPU is introduced by Zilog.
1975	The Altair 8800 computer is introduced by MITS.
1975	Microsoft is formed by Bill Gates and Paul Allen.
1975	The laser printer is introduced by IBM.
1976	The X.25 standard is introduced by the ITU.
1976	The Cray 1 is introduced.
1977	Apple II is introduced by Steve Wozniak and Steve Jobs.
1978	5.24-in floppy drive is invented by Radio Shack and Apple with Shugart Industries.
1979	The Motorola 68000 CPU is introduced.
1980	The Seagate hard drive is introduced by Shugart and Seagate.
1980	Optical WORM storage is introduced by Phillips.
1981	The Osborne 1 is introduced by Adam Osborne.

1981	The IBM PC is introduced.
1981	MS-DOS is introduced by Microsoft.
1984	The Macintosh computer is introduced by Apple.
1984	3.5-in floppy drives are introduced by Sony.
1985	The CD-DROM is invented.
1985	The Intel 80386 is introduced.
1991	The Intel 80486 is introduced.
1992	Apple and IBM join forces.
1993	The Information Super Highway is born.

HISTORY OF TELEPHONY

1851	First telegraph cable is laid under the English Channel.
1856	Western Union is founded.
1860	Philip Reis transmits sound electrically.
1875	Alexander Graham Bell invents the receiver.
1876	Alexander Graham Bell files a patent for his telephone.
1876	Elisha Gray files a patent for his telephone hours after Bell did.
1876	Bell makes famous "Watson, come here" transmission.
1877	Bell receives his second patent for an improved telephone device.
1877	The first permanent telephone wire is installed.
1877	Bell Telephone Company is founded.
1877	Bell marries Mabel Hubbard on July 11.
1877	American Speaking Telephone Company is founded.
1877	Thomas Edison invents the carbon transmitter.
1878	The first switchboard installed in New Haven, Connecticut.
1878	First operators were hired (all males).

1878	The first woman operator was hired.
1879	Bell resigns the board of Bell Telephone Company.
1879	Telephone numbers are used for the first time.
1880	Bell resigns from the Bell Telephone Company as Chief Engineer.
1881	Western Electric Company is purchased by Bell Telephone Company.
1882	New York is serviced by five telephone companies.
1885	Indiana begins deregulation of the telephone industry.
1891	Almon B. Strowger invents the step-by-step telephone switch.
1896	Almon B. Strowger invents the rotary dial.
1900	American Telephone and Telegraph (AT&T) is founded.
1907	Theodore N. Vail is named president of AT&T.
1913	The Kingsbury Commitment was signed.
1913	The crossbar switch was invented.
1915	The first wireless telephone call was placed.
1915	The first transcontinental telephone call was placed.
1920	AT&T entered into radio broadcasting.
1921	The first step-by-step switch was installed in a central office by Bell Telephone.
1922	Alexander Graham Bell dies.
1923	A short-distance telephony submarine cable is laid between Los Angeles and Catalina Island.
1925	Bell Laboratories is formed.
1925	AT&T sells all international plants to ITT.
1926	Bell Telephone introduces the first synchronous-sound motion picture.
1927	Time of day service begins.
1928	Major studios sign a licensing agreement with Western Electric for sound.

1930	Bell Telephone purchases a working crossbar switch from Europe.
1934	Watson, Bell's assistant, dies.
1934	The Communications Act of 1934 signed, forming the FCC.
1936	The first coaxial cable is put into service.
1938	The first crossbar switch is installed in Brooklyn, New York.
1948	The first No. 5 crossbar switch is installed.
1949	The Government Antitrust Suit is filed against Bell Telephone.
1950	Microwave is used for the first time in telephony.
1951	Direct distance dialing (DDD) is offered commercially.
1956	Transatlantic-oceanic cable is laid.
1956	The AT&T Consent Decree is signed.
1960	There is a field trial of electronic switch in Morris, Illinois.
1962	Telstar, the first international communications satellite is launched.
1965	Earth satellite is available for extending telephone service around the world.
1965	The first ESS office is in service in Trenton, New Jersey.
1968	The Carterfone Decision is signed.
1969	The MCI Decision is signed.
1970	Picturephone service is offered to the public.
1974	A government antitrust suit is filed against Bell Telephone.
1984	Divestiture of the Bell System.
1995	FCC begins auctions for PCS frequency blocks and markets.

1996	AT&T splits into three different companies.
1996	A new telecommunications bill is passed, allowing local and long distance carriers to cross into each other's territories.
1996	The Consent Decree is made obsolete by the Telecommunications Bill of 1996.
1996	Southwestern Bell acquires Pacific Telesis.
1997	Worldcom acquires long distance giant MCI.
1997	Bell Atlantic acquires NYNEX.
1998	Southwestern Bell acquires SNET.
1998	Southwestern Bell announces plans to acquire Ameritech.
1998	AT&T jumps into merger game, acquiring Tele-Communications Inc. (TCI).
1998	Cable & Wireless acquires MCI's Internet backbone (including UUNET).
1998	Bell Atlantic acquires GTE.

Index

ABOUT THE AUTHOR

Travis Russell is Regional Sales Manager of the Network Switching Division at Tekelec in North Carolina. He has been a field engineer in the telecommunications business for over 20 years, and lectures on basic telecommunications at area colleges and universities, as well as industry seminars. He is the author of *Telecommunications Protocols* and *Signaling System #7*, and the co-author of *CDPD: Cellular Digital Packet Data Standards and Technology*, all published by McGraw-Hill.